Drug Discovery

from Bedside to Wall Street

by

Tamas Bartfai, Ph.D.

&

Graham V. Lees, Ph.D.

AMSTERDAM • BOSTON • HEIDELBERG • LONDON
NEW YORK • OXFORD • PARIS • SAN DIEGO
SAN FRANCISCO • SINGAPORE • SYDNEY • TOKYO

Academic Press is an imprint of Elsevier

Elsevier Academic Press
30 Corporate Drive, Suite 400, Burlington, MA 01803, USA
525 B Street, Suite 1900, San Diego, California 92101-4495, USA
84 Theobald's Road, London WC1X 8RR, UK

This book is printed on acid-free paper. ♾

Library of Congress Cataloging-in-Publication Data

British Library Cataloguing in Publication Data
A catalogue record for this book is available from the British Library

ISBN 13: 978-0-12-369533-8
ISBN 10: 0-12-369533-3

For all information on all Elsevier Academic Press publications
visit our Web site at www.books.elsevier.com

Printed in the United States of America
05 06 07 08 09 10 9 8 7 6 5 4 3 2 1

Legal Disclaimer

CONTENTS

PREFACE

The pharmaceutical industry attracts daily interest. These are some headlines[1] during a not totally atypical month between January and February 2005.

"Painkiller (Vioxx) linked to 140,000 heart attacks in patients (of whom 44% died)."

"Merck is to spend $675 million on legal fees to defend itself against writs from patients who suffered side-effects from the arthritis drug Vioxx."

"Pfizer defends and continues promoting Celebrex (its own Vioxx) for current use and for potential new uses in cancer trials."

"Shares in Elan and Biogen Idec plunged as the firms suspended sales of new multiple sclerosis drug Tysabri after a patient's death (finally 3 died) in the US."

"AstraZeneca saw a bad end to 2004: a patient on its cholesterol-lowering medicine Crestor died and shares fell 2.5% before recovering. Two other drugs, Exanta and Iressa have also hit problems."[2]

"Novartis has moved into generic drugs with the $8.8 billion acquisition of copycat drug manufacturers Hexal and Eon Labs."

"Pfizer and Microsoft joined forces to file 17 lawsuits aimed at cracking down on email spammers selling cheap generic versions of Viagra."

"GlaxoSmithKline is facing a demand for $2.6 billion in backdated tax and interest on top of an existing $5.2 billion claim from the US taxman, enough almost to wipe out a year's profits at the UK's biggest drugs manufacturer".

---

[1] Largely taken but edited from British sources: *The Guardian Weekly*, the BBC and *The Independent*. The American Press has even more coverage of these stories.

[2] In order to assuage its investors, AZ promised to give them higher dividends.

"Rotavirus is the leading cause of severe diarrhoea in infants and young children worldwide and every year it kills 600,000 children below the age of five. A new vaccine for rotavirus is about to be licensed, in the hope of saving tens of thousand of young lives."

"Bayer," which following its voluntary withdrawal of Baycol was in purgatory, "are now the investor's darling because they have down-sized," i.e., people lost their jobs, "and have focused now on cardiovascular control and cancer drugs," which roughly translated means that other potentially promising lines of research have been shut down.

"Vioxx could soon return to market after a panel of regulatory experts in the US ruled that the benefits outweigh the risk." And on the same day "US curbs class action suits," and "Bush administration introduces legislation to curb frivolous class action lawsuits."[3]

A few months later we would have to focus on Viagra. By the time you are reading this there will be more such headlines, we are sure.

This tells any objective person that the drug industry is a big business with many "feeding at the trough." It is easy to attack the industry, but it is difficult for the industry to fight back in a reactive way. All responses have to be measured. Certain lawyers, representing patients but often to great personal benefit, are not seemingly constrained at all. In the United States at least some lawyers seem to say whatever they like to preserve the idea that the company is guilty.

The real problem for society is that the parties in any dispute about a drug are not trying to be reasonable to protect the public. Pharmaceutical companies are trying to sell medicines. If a medicine turns out to be "bad" for certain people to take, then the companies are accused of trying too hard to sell it. Any executive providing any off-the-cuff information about a drug that unwittingly reinforced the idea that taking drugs has an inherent risk would likely be fired, even though everyone should be aware of the drug's dangers. Going through the courts does not get you closer to the truth. Both parties adopt a position and defend it against all the opposition's attacks. A position is the worst thing to adopt in any negotiation.

[3] It was specifically thought (by a Democratic party spokesman) to protect "tobacco, oil, chemical and asbestos industries" rather than the pharmaceutical industry. This "intended protection of big business" may have little protective impact on the drug industry since any graduate law student would know that the death of patients as a result of taking a medicine is hardly frivolous. Time will tell, but this thrust of legislation isn't a license for the drug industry to act irresponsibly and, importantly, neither is the pharmaceutical industry viewing it as such.

But it's a sign of weakness to be reasonable. Objectivity is the first thing to go, and judges, trained as lawyers by years of practicing adopting positions on behalf of their clients, are suddenly supposed to be objective. Juries, often purposely selected because they have no relevant experience, have a choice of taking one side or the other. No matter how the evidence stacks up, it's win or lose. Whoever wins a particular case, society has lost. Either it has been taking an allegedly "dangerous" drug for too long, or it pays directly because the costs of lawsuits are transferred eventually to the consumer and indirectly because, as we shall discuss throughout this book, future drugs are dropped because of the "risk aversion" prevalent in the industry and increasingly also for the FDA. It is a matter of opinion if drug companies are paying for the lawyers' private jets or the consumer society is paying with higher drug prices.

On all of this crest of confusion, books emerge which always "have an angle." *Prozac Nation*—a very marketable title—is not about why far too many people started taking an antidepressant; it's about one person from an arguably dysfunctional family taking a mood stabilizer because it might (and did) help. Marcia Angell, who was editor-in-chief at the *New England Journal of Medicine*, authored a book entitled *The Truth about the Drug Companies: How They Deceive and Exploit Us, and What to Do About It*, which points out that drug companies spend a lot on advertising and do this with physicians acting as their agents.

A more profound "revelation" is that members especially of the U.S. society are prepared to take too many drugs with little provocation. Perhaps this is a little-cited observation of relevance. If you buy something simple like the anti-inflammatory and painkiller ibuprofen in the United States, you can get it from any supermarket in a multitude of brands and doses and formulations and, most pointedly, in bottles containing hundreds of tablets. If you go to a pharmacy in somewhere like Finland, you have a choice of about 2 brands, 2 doses, and 2 formulations and either a pack of 10 or 20. There are in total about 10 packets to choose from, and they are not in the supermarket; they are just in the pharmacies. Either living in America gives you more headaches, or the U.S. society has no fear of taking medicines. It's taken a lot more seriously in Europe. The most interesting question about Vioxx is not "Is it dangerous and did Merck know it was dangerous?" but "Why did so many people take it when they had so much ibuprofen in their medicine cabinet?" Vioxx was approved for arthritis and was taken for almost any pain or ailment; 80 million patients in 80 countries were allegedly taking Vioxx. The most interesting thing about Viagra according to the first headlines about it being linked to blindness is that patients noticed the symptoms and still kept taking it.

Problems of safety with drugs are not new. The machinery for tracking problems and for taking drug companies to task for problems is really new and supported by the Internet. If one looks beyond Vioxx to a much older drug—launched in 1989—used to treat patients with Parkinson's disease and restless legs syndrome called Permax,[4] one can find that this drug, after long-term use, can cause heart valve failure (similar to that caused by the weight-loss drug Fen-Phen). This discovery was first observed by physicians at the Mayo Clinic. Eli Lilly announced a safety alert in 2003. Are the lawyers as interested as they are in Vioxx? No. Permax is taken by fewer patients, has no off-label use, and this side effect hasn't been found in many patients. Cautious and vigilant physicians are most likely switching to alternatives. Of course, if you do web searches, you can find eloquent analyses of the problems associated with Permax on many lawyers' web sites, and they will counsel prospective clients about whether they might have a case and be able to sue. But unlike other cases (e.g., Vioxx and Baycol), the lawyers are not taking to the TV airwaves to find patients with cases for class–action law suits.

American society arguably treats pharmaceuticals as consumables, and the marketing and consumption of them follow similar patterns to those of soft drinks or fast food. Most people surely know that obesity especially in the United States can be correlated with the massive increase in consumption of, for example, high-fructose corn syrup, but the concern about obesity does not correlate with a decrease in its use in food manufacture. Drugs are treated as consumables by seller and buyer, and even prescription drugs are marketed in the United States in a way that prospective patients can assume they'll feel better and be better if they take them.

How can society come to understand that no drug is without risk? The risk may be small either because the side effect is definitely minor compared to the original ailment or because the chances of an individual experiencing the side effect is small. But there is always a risk.

What is the real issue? If you ask patients with HIV whether the drug industry has shown itself to be "their savior" and if they're rich or insured and living in the North western Hemisphere, they may say "yes." An infection—HIV—that led to a disease—AIDS—that was thought to be 100% fatal 15 years ago is now controllable thanks to a cocktail of drugs from three companies. However, if the infected individuals live in a developing country, the answer will not be the same. If they know about the disease and the drugs, which in itself is a big if, they will often be denied

---

[4] Generic name: pergolide mesylate.

access because they or their society cannot pay for it.[5] Their saviors may seem to be the company—maybe in Brazil or India and often in violation of a patent in some other territory—which copies the drugs and makes a more affordable version. Is that a problem of the drug companies that spent hundreds of millions or billions of their own money and are required by the rules of the market economy to make money, or is it a political and societal problem?

The questions "Why are new drugs so expensive?" and "Why I cannot afford them?" are the same.

Throughout this book, we will try to remain unbiased because, quite seriously and quite unfashionably, we are not on any side. We cannot argue against a capitalist system that has thrown up all the drugs in the pharmacy over any socialist system that provided no drugs (but that might do a better job of vaccinating its populations). But we can reveal a little frustration at the real root causes of all the problems the industry and society face. There are simply not enough affordable drugs for enough diseases. Why not? Well if you need to have a single group to blame, then we offer two: the lawyers who litigate and the venture capitalists who may want too much return from too short an investment and can switch their investments and allegiances on a whim.

The motivation for writing the book comes from the surprising observation that, despite the fantastic interest in the pharmaceutical industry, it is amazing how many people misunderstand it. This book is for everyone with a passing or professional interest in the amelioration of disease through pharmaceutical therapy. If you have expectations from, or suspicions about, the drug industry, then this book is for you.

Everyone expects something from the drug industry. Physicians and patients, investors, regulators, and administrators all have an active interest. Everyone wants to know what makes drugs "work" medically and economically. Why are drugs so expensive? What governs the pharmacoeconomics? Why are so few diseases treatable? The book is written in order to engage these audiences of varied backgrounds and vocabulary. Our didactic style uses examples. Some examples are intentionally repeated in different sections or chapters within a section simply to reinforce the framework of this complicated story. We wanted readers with no formal pharmacological training and who were approaching the subject for the

[5] A cynical but possibly valid view of the Bush administration's donation to fight AIDS was purely a way of getting U.S. drugs into the developing market (i.e., the money was intended to pay for U.S. drugs and go to the drug manufacturers rather than directly to the affected countries and patients to spend as they saw fit). The policy might also increase sales for the Swiss and UK manufacturers which make anti–HIV drugs.

first time to gain comfort from the emerging familiarity of the examples. We apologize for those readers who, being at least partially familiar with the material especially the scientific examples, find some repetitions irksome. We also do not expect every reader to read every page. It is not a textbook; the parts are written with different audiences in mind. Repetition, for emphasis, is largely intentional. Overall we hope this built-in redundancy is appreciated and effective. No matter from where the reader begins, we hope that he or she will be surprised about how many factors are involved.

This book is intended to open the windows and doors of the industry. It tells the story of drug development by using real stories about drug development from inside the process. The book will reveal negative aspects of the pharmaceutical industry, but simultaneously it may help build respect for the scientific and clinical efforts of the often maligned industry. Throughout, we have kept in mind general and specific misperceptions. As well as identifying many problems within the drug development process, the book offers solutions that extend well beyond the industry into society and legislation. The proposed solutions aim to make more and better drugs against more diseases accessible to more people. There are omissions, of course. For example, we haven't commented specifically on India joining the World Trade Organization and paying royalties on its previously low-cost HIV treatment cocktails.

We hope the book is a revelation even to scientists working in the industry. We hope business people in the industry develop a broader knowledge base for their decisions. We hope that politicians and lobbyists work more constructively when determining legislation, and do not polarize society. Legislation should facilitate the discovery and development of drugs. We hope physicians and their patients are rewarded with an improved quality of life through better medicines and treatment. A proper perspective on the industry is essential as society moves towards more legislation to control the drug industry without a full appreciation of the consequences of such legislation. While we advocate change, the authors are patently not activists.

If this book has the result of making more drugs available, more easily, more cheaply, for more disorders, in the context of a vibrant, science-driven industry, it would be a good result. That is our aim.

This book isn't about the stuff that makes headlines in national and international newspapers. It's about what really matters in drug discovery from the perspectives of patients, science, medicine, society, business, governors, investors, and, yes, even lawyers.

Tamas Bartfai, Ph.D.
Graham V. Lees, Ph.D.
San Diego, October 2005

ACKNOWLEDGMENTS

Regrettably, the authors have only themselves to blame for their mistakes. We thank in advance our scientific colleagues who understandingly forgive any mistakes made often in the interest of clarity—somewhat tarnished by exubelant style—over precision. We thank them in advance for correcting us.

We would like to acknowledge the unheralded contributions made by various mentors and sources, who over the years have imparted knowledge and wisdom to our absorbent brains. We have tried to acknowledge every source where possible; we apologize if due reference is seemingly not given.

We would like to thank the Skaggs Institute for Research and its President, Claudia Skaggs Luttrell, and Julius Rebek, the Director of The Skaggs Institute for Chemical Biology at The Scripps Research Institute, for being a catalyst to this book. Specifically, Tamas Bartfai thanks them for their encouraging him to organize a wealth of complex material for broad-based understanding, and Graham Lees thanks them for their general encouragement and actual financial support in forming this wealth of material into the book herewith.

Professors Peter Aranyi and Julius Rebek have provided both the scientific discussions and the friendship without which Tamas Bartfai does not function. He acknowledges their insights and thanks for sustained support.

Graham Lees would also like to thank Tamas Bartfai without whom the book would be empty.

About the Authors

Tamas Bartfai, Ph.D.

Professor Bartfai is Director of the Harold L. Dorris Neurological Research Institute, and Chair and Professor of Neuropharmacology at the Scripps Research Institute. He is a native Hungarian; he studied mathematics, physics, and chemistry in Budapest. He has a Ph.D. from Stockholm University in biochemistry in 1973 and was Professor of Neurochemistry and Neurotoxicology at the University of Stockholm and Professor of Medical Biochemistry and Biophysics at the Karolinska Institute. He has held positions at Yale, UCLA, and Rockefeller University as visiting professor. In 1997, he became Head of Central Nervous System Research at Hoffman La Roche in Basel as Senior Vice Director. He has to this point a 25-year career with 300 publications and dozens of patents in the field of physiological chemistry. He specializes in neuropeptides but embraces other broad areas such as fever.

Dr. Bartfai has been involved in the discovery and development of several drugs that affect the central nervous system (CNS). He has been involved in the development of drugs taken by more than 20 million people every day.

He is one of the few expert scientists who is equally at home in the boardroom and classroom.

After moving to San Diego, he became intrigued by the activity of his colleagues in the Biotech industry. But he was also simultaneously astonished by how little "academics" knew about drug discovery and development. This book is intended to remedy this situation.

Graham V. Lees, Ph.D.

Dr. Lees received his Ph.D. in neuroscience from the University of Cambridge in 1979 and performed postdoctoral studies at Gif-sur-Yvette, outside Paris. He embarked on a successful career in scientific publishing, holding senior positions at Elsevier (Amsterdam), Raven Press (New York), Academic Press (San Diego, New York, London), and The ScientificWorld (San Diego, Boynton Beach, Los Angeles, Newbury, Helsinki). Intrigued by the difficulties authors seemed to have in communicating complex ideas to broad audiences, he started writing and editing for several of the books he was publishing. He is still the Publishing Director for TheScientificWorld and an ad hoc consultant to scientific societies and the publishing industries. He is convinced that the successes of the drug development industry can be increased and its failures reduced. The success of this book will be measured in terms of improving the odds.

Disclaimer

Authors do not give investment advice and do not own a single stock in Biotech or the Pharma Industry, except some unrealized stock options and possibly via mutual funds. While the authors have made reasonable efforts to ensure the accuracy of the information herein, readers are advised to check other sources before acting on any of the information herein.

Introduction: Delivering on the Promise

It is very easy in these times to write about the promise of biotechnology ("Biotech") and the pharmaceutical industry ("Pharma" or "Big Pharma"), and to enthuse over what new biology ("genomics and proteomics," etc.) brings to the table. But the translation of promise into the business of treating diseases is complex. The inherent complexity means it is probably at least partially misunderstood by almost all the people involved in drug discovery: the academic scientists, all those who work in the industry, regulators, Wall Street and its international equivalents, economists, politicians, physicians, patients, their advocates, and society at large.

For an industry whose products many crave, Pharma is often maligned. Attacks come in all shapes and sizes. Drugs are "too expensive" or "not available" for enough disorders and complaints. For other industry watchers including those within the industry, risks are too great. Potentially good drugs are rejected for reasons that are scientifically not sound. If a drug causes harm, the entire company is at risk. Do such attacks on the pharmaceutical industry protect society or punish society?

Drugs are being introduced and withdrawn continuously. There is of course more written about those that were taken by many like painkillers and which, therefore, had sales in several billions of dollars—more people are affected and larger economical losses make for bigger headlines—than in the case of smaller drugs used by a few thousands of patients selling for, say, $50–200 million. The reasons for the withdrawal of small and big drugs are the same: our understanding of the risk-benefit ratio has changed since the drug was approved by FDA and introduced into the clinic, and many more patients have taken the drug for a longer time permitting us to discover more side effects. Vioxx is just one big drug withdrawn recently.

The recent withdrawal of Vioxx from the market presents a new backdrop to the industry. To understand this, it is necessary to understand some of the basic biology behind the family of drugs of which Vioxx is a member. It is a case of "better" drugs not being "better". Vioxx, Celebrex, and Bextra were never said to be more effective than aspirin, ibuprofen, or naproxen, but were sold based on that they were more "selective" (inhibited only enzyme COX-2) and not as the older painkillers both COX-1 and COX-2. Indeed this specificity translated into the expected reduction of risk of stomach bleedings, but as later trials showed it brought the unexpected increase in the risk of heart attacks. The "better" characteristics of the drug—that is, the fact that it was more specific than its competitors—was its downfall. The scientists can certainly be accused of making a mistake that was arguably predictable. Who was at fault? The scientists, the regulators, or the sales and marketing enthusiasts who pushed Vioxx on to such a broad public? The very failure of the drug should make society and business question their motives. How did Vioxx come to be taken by so many people? The demand—from both patients for drugs and investors for profit—is a significant part of the problem. Is capitalism in the dock alongside Merck? Would society benefit if Merck went bust? We address all this from scientific, medical, and business perspectives.

How can anything go wrong in the context of drug development being the most regulated human activity? The very consequences of this regulation are poorly understood. Decisions made behind closed doors in the pharmaceutical industry can determine whether drugs are developed or even whether diseases are considered for treatment. Drugs may be "killed" and diseases rejected for reasons not related to their efficacy or seriousness. Drug development is also a huge industrial enterprise with enormous economic weight.

By giving a variety of perspectives throughout the book, even patient scientists working in an academic setting and with at least one foot in the biotech industry should experience an epiphany while reading further. A fundamental issue is how better to translate basic science into drug discovery, and how to improve and even enjoy the process a little more. Many accomplished scientific colleagues are dumbfounded by the industry and by the decisions they observe as consultants to different Pharma[1]

[1] "Pharma" is often used in the book as an abbreviation for "the pharmaceutical industry."

projects. They may have just made the perfect compound. The compound is exactly as requested, and the project is suddenly and irreversibly stopped. Being scientists and being brought up in the tradition that you want a rational answer—with all the caveats attached to it—they cannot understand. "How is it possible?" "How are these decisions made when such intellectual value has been created, at such a cost, over such a long period with everyone's approval and then suddenly the project is no more?"

The truth is that despite the common education of scientists who are active in life sciences in academia and the Pharma industry, the values and work forms diverge very rapidly. And if academic scientists understood the industry a little bit better, things would be easier to comprehend and to accept. Increased understanding of the drug discovery process improves scientists' efficiency in translating scientific results into therapies.

The book addresses not only the general perspective of the Pharma industry and introduces drug discovery as a major science-based activity, but it also delves into the whole new set of values that come into play when a drug is ready to be tested in humans.

And, just when it became clear, the Pharma industry restructures itself for many reasons, not all good. There is constant revision of the business landscape. Is there motive behind the reason? The reason is "business needs," not just to make it more difficult to comprehend. But from the outside it seems that the geological parallel of the Pharma industry would be Mount Etna during an earthquake.

part
I
MEDICINES FOR SOCIETY

chapter

1

THE ART OF PUTTING A MOLECULE INTO MAN

THE PHARMACEUTICAL INDUSTRY: PERCEPTIONS AND MISPERCEPTIONS, PROFITS AND LOSSES

For an industry that attracts so much attention, it is very surprising how little is known about its inner workings. Conflicting views and opinions permeate industry narrative.

The ways of the industry might seem mysterious, but they are not. They are just complex, nonlinear, and based on several different value systems.

The Pharma industry is an old industry. It grew out of the recognition that making dyes for coloring uniforms is a great business but not as great as making drugs. To be more precise, the beginnings of the Pharma industry can be traced back not so much to a great wish to eliminate human suffering, but to give added value to a fine chemical company's products. Companies in the dye industry, such as BASF (Badische Anilin- & Soda-Fabrik AG founded in 1865), produced aniline and other indigo derivatives to make blue uniforms[2] without the need to import from British–controlled sources. Aniline derivatives were then developed as drugs for diverse diseases.

---

[2] In which one was a clearer target at which to shoot.

Both critics and advocates see the drug industry as a very large economic activity, which it most certainly is. Drug development provides the highest added value to products in the chemical industry and it is a profit-driven activity. It is really an industry that gives "added value." If you think about the chemical raw materials it starts with, they cost "peanuts." And, through the ingenuity of biologists and chemists, we make them into very valuable compounds. The industry, with annual revenues of some $213 billion,[3] is valued at $2.8 trillion.[4] It is also a very important commercial and societal force. In the United States alone it employs some 400,000 people.

Society should realize that there are considerable benefits to having scientists working on drugs to provide pharmacological therapies for old and new acute and chronic diseases. Scientists, in turn, should realize that there are few more rewarding moments than realizing that pieces of information that you have revealed may contribute to the development of drugs. The closest one can become to be a healer is to create effective medicines. A good psychoanalyst can help 120 people in his or her lifetime owing to the slowness and individual focus of the method, and, hopefully, those each interact with a hundred others. It's a very small number of people helped, no matter how good a psychoanalyst he or she is. If you make a successful drug, however, you may help millions and millions of people for decades until better medicine replaces it or the disease is eradicated.

A Strong, Regulated Dose of Safety

Many are eager to produce new medicines. This process, however, is the most highly regulated human activity. Which adverse effects a drug may cause in the large patient population during the decades of its use is a significant concern of governments, as well as of the pharmaceutical companies and their researchers. Indeed, governmental regulations are geared much more toward safety than efficacy of drugs. Safety is a tremendously important discussion in every Big Pharma company. Biotechs often come to realize this too late. They have often set up their entire enterprise upon efficacy of the novel treatment their research has produced.

A "safe drug" is **not** a "drug that won't hurt anybody." There are no such drugs. **Every drug can be taken against recommendations in high doses where it will become harmful**. And even at recommended doses most

[3] This figure is from the end of 2003 specified by PhRMA (www.phrma.org), an organization representing and serving as the mouthpiece for the pharmaceutical industry.

[4] This figure is derived from June 2004 data available from Biospace (http://www.biospace.com/stocksort.cfm).

drugs have some well catalogued—and thus for physicians known or expected, yet for patients only later apparent—unfortunate side effects. A **safe drug** is one for which we can predict these side effects, and where we can take measures[5] to predict who will be at risk, and take this person off the drug if this risk is life-threatening or very serious. That is a safe drug. If a company has developed a drug with which nobody has had any recognized problem within years of its trials and clinical use and then, all of a sudden, four people using the drug die, it is an "unsafe drug." It doesn't matter that it had been used without incident in, say, 5 million people before. If we don't understand why these four people in particular died, then they are reported as "idiosyncratic deaths while on this drug." That is ***not*** a safe drug. It will be withdrawn by the company, or physicians, fearing the worst, will not put new patients on it. If alternative medicine is available, physicians will also move the patients who presently do well on this "dangerous drug" to those alternatives. In another case, if it is found, using a new and simple laboratory test that only 140,000 people out of 200,000 can remain on a drug and 60,000 have had to be taken off the drug, it may or can still be a safe drug for these patients.

Later, we will discuss many cases of how safe drugs have been used to treat conditions for which they were originally not intended. Clinicians bravely try safe drugs in underserved indications like pain where they find that their arsenal is inadequate to assist their patients fully. "Safe drugs will find a condition to treat" is an old maxim. That is not to say that a company should—or is legally allowed to—promote so-called off-label use. However, it is a free and commonly followed practice in designed clinical trials to try old drugs for new indications. This *praxis* often brings the first drug to a disease thought to be too small economically or for which it was too difficult to develop a drug. This praxis benefits patients and companies alike when successfully put into practice.

For the drug industry, safety assessment is paramount. Risk aversion has a major effect on why drugs often do not reach the marketplace. It is safer to be extremely cautious than extremely brave.

Why? The main unstated reason is that companies are sued against their worth, not the extent of damage caused by use of the drug in question. The second, more palatable reason is that the drug industry is heavily regulated by governmental agencies like the U.S. Food and Drug Administration (FDA) and its international counterparts.[6]

---

[5] For example, by following a patient's enzyme activity and serum levels of other particular indicative measures.

[6] Such as the European Medicines Evaluation Agency (EMEA), or the British Medicines Control Agency (MCA) and Committee on Safety of Medicines (CSM), or Italy's Pharmaceutical Commission. Throughout the book, general references to the FDA imply "or other international regulatory body."

How is drug discovery the most regulated of all human activities? If you have the funds to launch a rocket to explore Venus, for example, it is actually not regulated, and as long as you find a somewhat partially regulated country that will let you put up your launch pad and you can hire the people who have the technology, you are fine. This is only a question of money; it is almost totally unregulated. But if you want to put aspirin and some other approved drug, in combination, into a patient as a study, it's regulated. If you want to put something that has never been in humans into a healthy volunteer, that's incredibly regulated. Is this bad? No. The Food and Drug Administration (FDA) and its international equivalents are there to protect the public. And they protect the public usually in a rather conservative but very efficient manner. It is not their job to invent new drugs; it is not even their job to tell the drug companies how to invent new drugs. It is their job to tell the drug companies that they cannot endanger people while testing or selling drugs. Once these drugs are approved and made available to physicians and their patients, it will then become a discussion of "therapeutic ratios"—that is, benefits versus drawbacks—between doctors and patients. It is the job of the FDA to follow what happens with every medicine marketed to be used in any patient in the United States. The FDA makes sure that companies collect and speedily report side effects encountered by patients taking their drug so that proper warnings can rapidly be issued to avoid further harm; controls that companies do not change advertisements and use warning labels where necessary; maintains safe production of approved drugs, and restricts the marketing of drugs to those diseases they were approved for, and only make promises and claims that are scientifically defendable.

There is no other more regulated form of human activity. Even with all the new campaign finance reform, politics is much less regulated.

Companies' Failed Drugs Have Lots of Company

Why is it that fewer than 10% of molecules which make it to the stage of being "put into man" in clinical trials become drugs? Why do only one or two of hundreds of Pharma companies' research projects produce drugs? Candidate drugs fail on either safety, toxicity, or through interactions with other drugs, or, less often, the lack of efficacy: they do not work well enough. Although toxicity can be better predicted now, animal models and electronic "expert systems" are far from perfect and will not for the foreseeable future replace healthy volunteers in testing drugs for safety; without volunteers there would be no new drugs.

What sort of toxicity? In some cases it can seem extreme, such as causing developmental disorders. Such so-called teratogenicity is more common than appreciated. For example, the anticonvulsants carbamazepine, phenytoin, valproate (also a mood stabilizer), the anticoagulant warfarin, the acne medicine isotretinoin, and some tetracycline antibiotics can all be made to show teratogenicity at certain (high) doses in animals. Probably the most infamous teratogenic drug, thalidomide, was prescribed in Europe in the 1960s as a sleeping aid and morning sickness pill. But its widespread use resulted in tremendous human tragedy from its causing drastically abnormal limb growth *in utero*. Meanwhile, in the United States the FDA was still studying it, saving the U.S. pregnant mothers and the whole society from a similar tragedy. This drug has now found a new use in the treatment of leprosy, a terrible disease without many treatments being developed directly, because leprosy is a disease of the poor world.

The crucial point is that armed with this cautionary information, physicians do not prescribe any potentially teratogenic drugs where the risk of birth defects would outweigh any benefits.

The message is that during the drug development and clinical trial process that takes 5 to 12 years, many potentially promising drugs are lost by being found to be not safe, not efficacious, or neither. And even for prescribed drugs, physicians and patients should without fail check the labeling for contraindications.

Patented Drugs vs. Generic Drugs: On the Table and Over the Counter

Our assumption is that some of this book's audience might be against "patenting" drugs. The argument is that generic drugs are cheaper; therefore, we should have free access to generic drugs. Unfortunately, the argument is perilous.

Patents are an extremely important aspect of drug discovery. There are examples of drugs being developed without patents decades ago—such as Tylenol (acetaminophen)—but if a scientist has made a major discovery that might have the potential, say, to alleviate pain, and doesn't patent it, it most likely will not be developed in today's world. Quite simply, nobody will put in the money to develop something around it. Scientists have an obligation to patent their discoveries. It is a real intellectual duty in our real world to patent what you have discovered or made. Scientists might even be accused of neglect if they forget to patent, because it prevents anyone from commercially[7] using their results (*for the broader benefit of humankind*).

The patent buys time and exclusivity. The economic model of drug research, discovery, and development needs patent protection if the discoveries are going to be economically viable. The original patent lasts 20 years from published discovery but can usually be effectively extended by 2 or 3 years. This is germane to the pharmaceutical business.

Occasionally, discoveries cannot be patented. For example, if you discover something novel, such as lithium ameliorating bipolar disorder, there is a problem. Lithium is a chemical element, not a new chemical compound that can be patented per se. Since the industry could not make much money making lithium tablets, there was less incentive to establish a use patent. Nowadays, for some cases such as this, there is an orphan drug statute in the United States, which, since 1983, gives a company a seven-year exclusivity and enables it to develop a drug treatment and still help patients. But this is a very slow and very difficult process. It is much better to patent, and let the patent drop should no one develop it into a commercial product, than not to patent at all.

A consequence of the protection of a company's business by establishing and enforcing patents is often seen as providing the opportunity to charge very high prices. This is a gross oversimplification. High prices are as much a consequence of the difficulty and high cost of finding and developing drugs, the high costs of marketing drugs, and the potential high cost of failure of the drug itself or others in the company's portfolio. The prices are to cover the costs and risks of research and development, marketing, the costs of capital during a long development process, and the cost of surveillance and marketing in a discriminating crowded market. The realization that a larger Pharma company must have the capital to develop simultaneously 30 products in expensive clinical trials costing up to $200–800 million[8] just to get at most three new medicines approved a year, indicates that this industry is more capital intensive than developing airplanes or computer chips to mention two other high-tech activities.

---

[7] The emphasis here is on commercial development since few medicines have yet been developed by charitable organizations, although many work hard on getting the funds and skills to do so. These would include foundations for cystic fibrosis, multiple sclerosis, and the Hereditary Disease Association for Alzheimer and Huntington Diseases.

[8] Drug companies typically spend 10–14% of gross revenues on R&D. Novartis, for example, may spend $1.6 billion on preclinical and $3.4 billion on clinical (2003 figures). Similarly, Bayer may spend similar ratios on preclinical and clinical such as $700 million and $1.3 billion on preclinical and clinical respectively (2003 figures). Depending on how costs are assigned within a large company—especially costs for programs which are stopped—annual expenditures across the whole program of multiple drug candidates at different stages of development vary between $200 million per year per clinical candidate and $1 billion per year per launched product. However, mergers are tending to reduce the percentage spent on R&D.

Are generic drugs precisely what the doctor ordered? No. Without patented drugs establishing the market, generic drugs would simply not be developed. In addition, the original formulation may deserve continued sales after the patent has expired since it is sometimes true that generic drugs may have a different formulation of the same active ingredient that may not be entirely inert and side effects have been reported. Thus, generics may not always be fully equivalent in every sense. The FDA checks that they are pharmacologically equivalent to the extent that they contain the same dose of the active ingredient, but formulation tricks and other minor but not totally unimportant things may not be, and do not have to be, the same. Remember that not all margaritas have the same color or are served in the same glass even if they contain the same amount of alcohol. Some are willing to pay for the difference. It may be unfair to adopt legislation against physicians prescribing the original formulation. Although insurers in the United States and whole national health care systems in such countries as the United Kingdom, Sweden, and Germany, recommend their physicians first to prescribe generic equivalent drugs to cut overall costs of the whole system, if the physician can justify why he or she prefers the original drug, he or she can insist on it even within these different systems.

This does not mean that no improvements to pricing policies of the pharmaceutical companies could be made. Indeed, radical beneficial changes could be made without jeopardizing the companies' viabilities.

Who Does What in the Industry?

First of all, in drug discovery there are two main players: "Big Pharma" and "Little Biotech."[9]

The original source of ideas for Big Pharma's products was its own research. This was until governments during and then after World War II started to outspend the basic research programs of Pharma companies. By 1960, the majority of research results were coming from academic research at universities. Between 1960 and 1980, pharmaceutical companies hired their own biochemists and, later, molecular biologists to discover biological processes and targets for their own candidate drugs. More recently, since the mid-1980s, a major source of the biological "targets" the drugs affect is Biotech. Incidentally, Biotech companies often choose to patent not only the candidate drugs, but also their targets, for sometimes

[9] Do not think of Amgen, Chiron and Genentech and their like as Biotech as these Biotechs became rich enough to work like Big Pharma.

it's the only value these companies will have. Big Pharma companies, on the other hand, often only patent the chemical entity that might become a drug. The philosophy is: "We are in the business of making high-value medicinal molecules; it's the molecule that we want to hold the patent on." For small companies, it's different. They often started with biological discovery from which they have discovered the target for which a novel molecule should be developed. If they get the human and financial capital, they may try to "go it alone," but in the overwhelming majority of cases they have to team up with Big Pharma for chemical know-how and for funds for clinical trials, and to tap into the decades of expertise.

The relationship between Big Pharma and Little Biotech is similar to the relationship between the Big Studios and Independent film companies. The "Indies" often have the ideas, the stories, the production, and so on, but they need the marketing muscle of the Big Studios for effective distribution. Incidentally, the film industry probably envies the shelf life of successful drugs.

There has been and will continue to be innovative reconstruction of the pharmaceutical industry with, for example, cross-licensing of products and expertise. Sandoz[10] collaborated with Bristol Myers Squibb (BMS) in oncology. Other companies have a more acquisitive outlook. Pfizer bought Warner–Lambert for its cholesterol-lowering Lipitor, and Pharmacia "merged" with UpJohn-Searle for its COX-2 inhibitor. Pfizer then acquired Pharmacia-UpJohn-Searle to become the world's largest pharmaceutical company.

Larger capital, research, and marketing resources emanating from these mergers are good news in this resource-intensive industry. At the same time, some specific negative consequences are emerging from the frenetic company mergers, and these will be examined later under business issues.

Within a Pharma company, successful drug development needs teamwork. We will deliberate on details later, but we have some preliminary thoughts by way of introduction.

No matter how good the basic science and the drug candidate are, and remember there are no perfect compounds—everything is poisonous at some dose and form[11]—if you do not have clever, courageous toxicologists with high personal integrity you might as well shut down the drug

---

[10] Subsequently integrated with Ciba-Geigy into Novartis.

[11] For example, OxyContin is a long-lasting slow-release—"depot"—painkiller, but when the ground tablet is snorted, it becomes a drug of abuse and can kill since too high a dose is released at once.

development operation. Biotechs don't seem fully to understand this. How would they? How many compounds have they put through the clinic? Two in their lifetime, maybe? So how would one know? We don't blame anybody, but what can be learned?

It is equally important that you need imaginative and courageous clinicians and volunteers to test your compounds in clinical trials. Why courageous? If "overly cautious" doctors test new drugs in mild cases—patients who would recover on their own—it is difficult to show a positive result, and the drug fails on lack of efficacy since those on placebo, being not so sick, also do well. "Daring" doctors may uncover great benefits of the drug in the trial but may also "kill the drug" by administering it to people who are too sick and where nothing would really help.

You must also have imaginative marketers, a whole army of them. A major Pharma company has as many as 10 to 15% of its workforce as marketing agents, specialized in such activities as selling drugs for cardiovascular or pulmonary disease.

An increasingly major role in executing clinical trials is being taken up by Clinical Research Organizations (CROs). For example, Quintiles, the largest with 2004 revenues of $1.8 billion, has, through its strategic partnering group PharmaBio Development, committed more than $1 billion to date to assist pharmaceutical and biotechnology companies and actually has invested more than $500 million to date. They have a vested interest in ensuring a steady flow of candidate drugs into their own clinical trials business. Incidentally, the CROs might be considered indifferent to clinical trial failures since they are paid according to the procedures agreed upon and the number of patients entered into the trial. For the success of the industry, however, it is important that they have very high standards when recruiting patients for the clinical trials and when conducting and evaluating the trial data.

Immediately outside the industry, waiting to administer and judge your results, is the FDA and international regulatory bodies. They determine the acceptable clinical endpoints to determine safety and benefit.

For a big drug company to fail to obtain approval of a particular drug may be "no big deal" because the company is working on, say, 200 to 300 projects and has at least 20 to 40 compounds in FDA review. If the FDA ruling causes a project to be abandoned, this may be terrible for the team that works on it, but for the company it's no big deal. For a Biotech company, which in its lifetime meets with the FDA three times at best, it's a big deal. Biotechs should always team up with people who have done this before, that is, taken drugs through clinical trials and regulatory approval processes. It is terrible if you have good science, and you mess it up on the trivial part of regulatory issues. It happens very often, and it's a tragedy because you might have defined a new scientific principle, a new target in a disease, but through butchering it with amateurism you may give this protein a bad name, and then nobody else will go to it. It is a tragedy for patients and scientists together.

The FDA approves drugs that have shown safety and efficacy during clinical trials. The agency does not require that the new drug is better than existing drugs, or is easier to administer. It is the companies' marketers who want to be able to make such statements. The FDA does not always require an already approved drug to be tested in parallel with the new drug; it is the companies' clinicians who want to know if the trial is properly conducted, and the way of showing that is by using drugs with known efficacy and demonstrating their efficacy in these trials too.

The FDA approves drugs faster when a new therapy or a significant improvement over existing therapy is presented. Copycat drugs with minimal differences may be approved, which allows more than one or two companies to make blockbuster drugs. For example, the FDA has approved no less than seven examples of a certain newer class of antidepressants known as SSRIs, each of which has a turnover of over a billion dollars.

The FDA also determines the labeling policy on drugs. The what, how, when to take it—indications—and the whom and when not to take it—contraindications—to be properly described, listed, and displayed. But the administration of actual prescriptions needs the physicians' expert control. They can choose a dosing different from that suggested by the trials and recommended by the company. They can also try the drug in an indication for which it has not been approved, or, indeed, for which it has not been systematically tested. The industry and society need responsible prescribing. Some drugs fail after approval when they are "postmarket." A bad label here—for example, one that cites a cancer risk—can kill a drug in practice even if it is formally FDA-approved.

chapter

2

RAISING AND RISING EXPECTATIONS

WHAT SOCIETY NEEDS FROM PHARMA

The expectation is that pharmaceutical companies should produce affordable drugs to cure all significant diseases. In reality, apart from bacterial infections, the industry ***cures*** no diseases; rather it only offers treatments—often for the symptoms, not the disease itself—for relatively few diseases, probably less than 8% according to the physiologist's definition i.e. when the patient returns to the exact same state of health prior to the disease. The fact that surgeons are regarded so highly means that the practice of medicine is not so terribly good. Surgeons are fantastic in what they do; they improve their techniques constantly. For example, in 2005 the best treatment for many cancers remains early and complete tumor removal followed by chemotherapy and radiation, which, in most cases, amounts to a "cure" without recurrence or worsening of the disease. In comparison, the pharmacological treatments are not that great, neither in tackling the bases of the diseases nor in curing them.

Furthermore, Pharma companies have not yet been very successful in treating the symptoms or in slowing the progression of many major diseases such as Alzheimer disease or osteoporosis. That being said, the slowing of disease progression can hardly be called an actual *improvement*, and it should not be the final goal of medical treatment. Best would be to cure diseases, or, at least, to stop them fully. That is not to say that slowing disease progression should not be agreed as a *sign* of drug efficacy and should not be pursued until we know how to stop the disease.

Society needs to have a broader commitment from the industry. It should expect that Pharma helps to cure a higher proportion of diseases and ameliorates more symptoms more effectively.

Society, comprising patients, physicians, taxpayers, investors, and government, expects more from the industry when it purchases its products at the rate of $300 billion per annum in the United States alone. The companies in turn spend $49.3 billion[12] of this revenue on research and development (R&D). On top of this, U.S. society currently spends $28 billion through funding of research at the National Institutes of Health (NIH). This investment should, and does, benefit Pharma directly.

Society should also expect that with better medicines some of the other costs of health care could be reduced. The $300 billion that society spends on drugs is only 14% of the roughly $3 trillion health care budget.[13]

One of the significant problems for the Pharma industry is that of the 400 disease entities identified only 50 are commercially attractive by today's requirements of return on investment (ROI). Society needs to find a way to make more diseases commercially attractive if it wants Pharma investment in treating any of the other 350 diseases affecting hundreds of millions of people.

There is another significant problem. In order to treat a disease, you have to have a biological target—usually a protein associated with a disease—for the drug to interact with. Today we have only about 200 clinically well-proven or validated drug targets, and most of our medicines target these. To make progress, we need to expand this number. New targets will enable the treatment of diseases that we haven't yet been able to treat, or make existing treatments better. One cannot possibly hope to treat over 400 diseases with only 200 or so clinically validated targets even when the same drug may often be used for several diseases.

This is difficult to legislate for, but society also wants Pharma to be more inventive in its approaches. If you ask pharmacists how many drugs there are, they'll probably guess about 10,000. There are actually well over 10,000 listed in the FDA's 1,000 plus page "orange book,"[14] which is the official source of available medicines. If you ask how many chemically fully different drug molecules there are, the most popular guess is usually

---

[12] In 2004; data from PhRMA (http://www.phrma.org/publications/publications//2005–03–17.1143.pdf).

[13] Estimate of global health care budget of which the U.S. component was about $1.3 trillion in 2003.

[14] 25th 2005 edition. http://www.fda.gov/cder/ob/default.htm

around 2,000. If you examine it closely, however, there are only 433 distinct chemical entities on the market in 2002. Most of the remaining 1,500 drugs sold are different combinations of the 433 basic molecules. For example, acetaminophen (Tylenol or paracetamol) is a component of 337 of the 10,000 plus drugs. Incidentally, acetaminophen is one of the more financially successful nonpatented drugs.

Consider also that for every really new drug there are "copies" coming out very rapidly, so-called me toos or followers. Since these are often in practice only slight modifications of the original molecules, the number of *truly* original medicines is really much less than 433.

Examining the chemistry even more closely, we see that these 433 entities are derived from very few basic structures, or so-called chemical scaffolds upon which classical pharmacology works. After 120 years of conscious, and perhaps cautious, medicinal chemistry and pharmacology, we have come up with very few new chemical scaffolds to interact with the biological targets.

Most chemists and pharmacologists find these data pretty shocking. But it is a worthwhile perspective for all of us to consider. We have so few drugs to show for so long and such expensive research not because we don't try hard enough, not because we are "idiots," but because it's extraordinarily hard to find effective, safe new drugs.

The picture painted here doesn't appear any rosier if you consider that most of the drugs in use as stand alone or in combination were approved before 1960 and, today, many of those would ***not*** be approved. Aspirin, for example, would not be approved today because of the frequency of stomach bleeding in its users. This would be true even if you also consider that we have come up with new applications—"indications"—for aspirin, such as to lower thrombosis risk by utilizing its recognized major side effect.

Concerned and informed members of society should also wish that Pharma would adopt more modest financial ambitions. The philosophy of always looking for blockbuster drugs that make billions of dollars per year on the market must change.

Pharma companies should also be more adept at being inventive. A clinical trial's design is too often the root cause of a drug's failure, and nine out of ten drugs that reach clinical trial, after five to eight years of preclinical work are failing as clinical medicines. We should hope for better selection of the patient groups likely to benefit from treatment.[15] We also need better methods for determining and evaluating a drug's efficacy during a trial.[16] Too often trials for complex diseases will take

[15] Genetical analysis or pharmacogenomics is much hoped to contribute to better stratification of patients who are likely to benefit from a given drug.

[16] For example, using surrogate markers.

too long. Does this mean that society would expect pharmaceutical companies to take more risks? Possibly! Would society welcome pharmaceutical companies taking more risks? Much less likely! For truly massive, long-term trials such as are needed to determine the risks and benefits of hormone replacement therapy for menopausal symptoms, the government sponsored the multiyear trial within the Women's Health Initiative. The increased cancer risk uncovered in 2002 led to a huge reevaluation of prescription practices of estrogen therapy.

But society should want companies to have less propensity to drop potentially viable candidate drugs. In fact, we see a renewed effort to "revive" failed drugs for new disease indications.

At the practical level, society needs more reliable, clearer, and more understandable information about the actions and effects of drugs, especially new drugs. Physicians need to be more involved in drug development and better educated about drug actions. Drug companies, especially sales reps needing sales, are often too keen to emphasize the benefits of a new drug; they seem to be less open about why and when physicians and patients might be better placed if they didn't prescribe and take it. Some of the information is confusing. Other information, such as that presented in the costly advertising campaigns aimed at potential patients, seems intrusive, invasive, and blatantly sales driven. Simply put, better labeling and more effective risk and benefit assessment together with, where appropriate, other "off-label" information, would be beneficial and in society's interest.[17] The FDA repeatedly has to warn Pharma companies about their advertising practices. There are no exceptions to this. Both Small and Big Pharma can overstate the benefits regularly and might neglect to list all known risks.

Various specialist groups within society are pushing the industry to fulfill their own objectives in the area of health care. Most obviously, patient advocacy groups want to see their particular patient population served with new, more efficacious treatments.

Business investors want more return from the companies. The Pharma industry is an important part of a healthy economy. It is financially appealing because it is potentially a constantly growing and noncyclical business. The return on investment over the last few decades has been without equal for any other major industry. Even though Pharma R&D is capital intensive, price to earnings ratio[18] returns for

---

[17] See, for example, http://www.ismp.org/MSAarticles/A2Q99Action.htm

[18] Calculated as the current market price of a company's common stock divided by that company's earnings per share in the previous 12-month period. See, for example, http://stocks.about.com/library/bldef_pe.htm

Pharma companies can be very high; so far, there is no shortage of investment capital, and the industry has been growing at a steady rate for an astonishing 50 years in a row!

What Pharma Needs from Society

If society wants Pharma to provide more treatments for more diseases, then it has to address the industry's problems.

Pharma used to spend 18 to 20% of its revenues on R&D, but as the revenues have grown, this percentage has fallen to its current 10 to 14%, not only because people became greedier for profits, but also because of the shortage of clinically proven targets. Throwing more money at the problem may not bring proportional rewards.

Pharma probably needs more cooperation from publicly funded research to find and elucidate targets. Too often, it seems that scientists engaged in pure academic research are not looking to help drug companies work with the targets they have discovered in the course of their research. They often zealously "protect" their data and do not always allow drug companies to have access to putative targets. The NIH and international research efforts should be providing Pharma with more targets. The Human Genome Project (HUGO) represents—albeit indirectly—the single largest such assistance from governments to the pharmaceutical industry and to basic academic research alike. Recent efforts in the "NIH Roadmap"[19] also aim at the NIH's more direct contribution to drug discovery.

With more drug targets and a better fundamental understanding of their involvement in disease, it is more likely that new chemical entities can be invented, tested, and brought into clinical practice.

How should society allow for new drugs with side effects, say, as "dangerous" as aspirin, to be approved in today's litigation-prone climate? Without resorting to stricter control of the prescribing of drugs, perhaps a solution might come through better clinical trial strategies aiming at smaller, better selected patient groups, and through approval of drugs that may not be useful for a significant proportion of sufferers because of contraindications but that may help others significantly. Not all drugs are for everyone, although it would simplify matters financially and clinically if they were. Pharma itself can determine this, providing it encourages responsible prescribing. In many cases, it has developed a very strong drug–drug interaction profile description to make safe prescribing easier for complex diseases.

[19] http://nihroadmap.nih.gov/

Pharma companies are not keen on taking risks; this is hardly surprising when they can be sued for all their worth. If something does go seriously wrong with a new drug, it is too easy and tempting to blame the participant with the deepest pockets. With all the consolidation through company mergers, risk aversion can be even more prevalent. The mergers themselves make the size of the putative defendant in any lawsuit much bigger; some of the companies are worth hundreds of billions of dollars. Other companies are left out from mergers because of pending lawsuits against them.[20]

Do Pharma companies need to be better protected from litigation? The answer might indeed be "yes." There are cases where drugs failed because of their prescriptive administration or self-inflicted overdosing rather than because the drug is ineffective or dangerous if prescribed and taken properly. Meanwhile, society itself has changed and accepts less risk. For example, the general vaccination programs introduced in the 1950s and 1960s such as the whooping cough vaccine with its serious side effects for 1 of every 3–10,000 children could not be repeated now. Yet an overwhelming majority of parents and physicians agree to use the vaccine to protect from this potentially fatal bacterial infection, and the Center for Disease Control (CDC)[21] recommends and supervises vaccination against whooping cough (See Box 2.1 for general background).

Box 2.1 General Vaccination Programs

Vaccination is about harnessing the great potential of the immune system to develop agents—antibodies—with exquisite selectivity. The oldest way of doing this is exposure to the infectious agent. People used to send their kids to a house with a kid with pox, to expose them, so that they would acquire immunity. Vaccination involves isolating and making less harmful the infectious pathogen, either by selecting a less virulent strain or otherwise weakening it and making an attenuated strain and then giving a controlled dose to healthy individuals. Ever since the Swedish Army was vaccinated 350 years ago, vaccinations have turned out to be the most effective and the cheapest means to save many lives.

However, despite the obvious importance of vaccines and society's reliance on them, there has been an overwhelming trend toward reduction in vaccine research and the number of vaccine manufacturers. Few drug shortages would have caused the political calamity that the 2005 influenza vaccine shortage in the

[20] Having pending lawsuits is indirectly a protective measure against hostile takeover.

[21] http://www.cdc.gov/

Box 2.1 General Vaccination Programs—*cont'd*

United States caused. Had the 2005 flu season been a severe one, it could have led to thousands of death in the United States. It is only luck, not political or commercial or medical wisdom, that prevented such a catastrophe. However, it is questionable if real conclusions are being drawn from this event.

There is a huge difference between vaccines and drugs. Drugs are developed to treat diseases and the manifested symptoms for which patients sought help. Drugs are given to diagnosed patients. Vaccines are given to healthy infants, children, and adults. They are not sick, they have no diagnosis, and as infants and children at a time when they do not have a well-developed immune system. However, successful vaccination programs effectively protect the entire generation of children, with 90 to 95% of an entire cohort being vaccinated.

Normal vaccination involves injecting or ingesting the antigen into the body and letting the immune response produce a series of antibodies over days. Boosting the antibody production with a second and third injection of the antigen is common in most vaccinations. There are heart- and lung-disease-afflicted people who cannot deal with exposure to the pathogen–antigen in the course of normal vaccination. For over a century, vaccinated soldiers gave their blood, from which the immunoglobulins were purified and administered to these individuals in a "passive immunization" procedure. Nowadays the neutralizing antibodies are produced in bacteria in fermenters. Passive immunization is common for snake venoms; when it is rare to be infected or injected with the pathogen, general vaccination makes little sense.

A subset of the population should be protected from infectious disease to a greater degree than society as a whole, as the members of this subset are more exposed. Police, border patrols, teachers, and medical staff will be more exposed than the population in general to infectious diseases that are brought by tourists and immigrants from other countries. Because of the success of vaccination and increased living standards, tuberculosis (TB) vaccination in most developed countries stopped in the past decades. This left the population totally unprotected when the fall of the Soviet Union opened the possibility for millions of people with TB—in fact often bacterial strains that were fully resistant to common antibodies—to visit or to come to work in Western Europe. The Eastern Bloc had general health care coverage that used antibiotics generously; this practice has led to the development of many antibiotics-resistant TB strains. TB is on the rise again globally, and thus new vaccines are needed; yet the few companies that still make vaccines are not very active. As with many new vaccines that are considered as a need for the developing world (i.e., a world that cannot pay), it is regrettably not considered a commercially worthwhile effort.[22]

continued

[22] See also Chapter 20.

Box 2.1 General Vaccination Programs—*cont'd*

Bacterial vaccines such as tetanus–diphtheria–pertussis (or triple vaccine) in most countries have been used for almost a hundred years. The biggest problem with these vaccines comes from some side effects of the pertussis vaccine. This has led to efforts to replace the whole bacteria-containing vaccine with a cellular vaccine, that is, where one or a few important proteins that are the major antigens inducing the protective antibodies are selected and administered instead of the whole-treated and formaldehyde-killed bacteria. It is clear that vaccine development today is not difficult; what is very difficult is the field-testing. In the Western world the only country in the early 1990s with pertussis epidemics was Sweden, which had suspended obligatory vaccination with pertussis because of complaints about the high frequency of neurological side effects such as persistent crying for one to two nights. Since pertussis can be treated with antibiotics in a highly developed country with good health care with rapid systemic antibiotics and rehydration administered in hospitals, the Swedish policy would not increase pertussis mortality. But suspending pertussis vaccination in the underdeveloped world would lead to 1 to 1.5 million deaths/year.

Smallpox vaccine was historically the first and arguably the most important viral vaccine. The eradication of smallpox has been among the greatest achievements of humankind and an example of what global efforts can achieve. The polio vaccine, which is 50 years old now, is an important example of what happens when development work on vaccines is stopped. The polio vaccine is an attenuated live viral vaccine, which gives a short-lived advantage to the body to make protective antibodies to fend off poliovirus infections. The problem with polio, as the Finnish polio outbreak of 1984–1985 showed, is that we do not develop new vaccines, although the pathogens mutate, and in many cases, we stop vaccination because we believe that we have eradicated a disease when the reservoirs still exist for the pathogen. Of course, by general vaccination programs the likelihood of epidemic outbreak is greatly reduced, and the individual's ability to withstand viral infection is greatly enhanced by the vaccination. Thus, there is a common societal interest in keeping vaccination rates high in spite of complacency, religious objections, and the like. The problem is when a new mutation occurs that produces a virus against which the vaccine strains' inoculation has not prepared us. With influenza vaccines we have accepted this possibility. We live with the understanding that each year's flu vaccine has to be made from that year's "crop" of flu virus, identified when it arises, usually in East Asia, and then cultivated and padded and administered, all in c.3–4 months. This is as seasonal an activity as picking tomatoes.

With a greater immunocompromised population caused by HIV, transplantation medicines, and other immunosuppressive treatments for rheumatoid arthritis (RA), some vaccinations utilizing "live vaccines" such as the old smallpox vaccine, can be very risky. Society now unwittingly tolerates the risk of a greater number of deaths through the infectious pathogen itself with this policy of nonintervention than it did using vaccines under a policy of intervention in the name of the "greater good."

In addition, the industry is loath to embark on programs for certain disorders, such as sepsis and stroke, where one of the outcomes of the clinical trial, if the drug is even partially ineffective, might be death. The industry generally avoids clinical trials where the outcome is measured in mortality. It just doesn't "look good" In addition, of course, a sepsis or stroke drug would only be administered for 1–7 days whereas a good antidepressant prescription would be for 6–12 months and ones for RA or osteoporosis, might be for 30 years, making them much bigger businesses. Yet we need drugs to save lives from deadly diseases. When the science and the courage are there, important reductions in mortality can be achieved as shown in the cases of using tPA and urokinases to treat heart attacks.

Pharma may on occasion need more volunteers. There is no new medicine without volunteers for clinical trials, and no trials are without risk.

It needs compassionate, courageous, scientifically well-trained clinicians to do clinical trials. Not all physicians are willing to take part in a trial because there are many potential risks in agreeing to be involved and few, if any, benefits except the rare outcome of the physician then becoming an opinion leader in his or her field of medicine.

It is hard now to imagine the courage of Charles Herbert Best who injected insulin[23] into a 14-year-old boy in 1922. He didn't hesitate or have to defer to any ethical committee, a 1960s invention. And insulin is still being used in substantially the same form he injected. As long as a physician believes that this will do good, and the patient agrees, or some consent is formed, it can be really worth doing. The courage of the patient, Leonard Thompson, should also be recognized. He was rewarded with a life extended by some 13 years. He died at age 27.[24]

If society expects Pharma to provide treatments for the more complex, difficult and less potentially financially rewarding diseases, then, with proper constraints and limitations of access and use, society needs to allow better access to patient information. One can in the near future also contemplate *individualized medicine.* This idea is very attractive and at the same time controversial since it requires that the individual genetic

[23] Insulin was discovered at the University of Toronto, Canada, in the summer of 1921 by Frederick Grant Banting and Charles Herbert Best. (http://www.discoveryofinsulin. com/Introduction.htm)

[24] http://eir.library.utoronto.ca/insulin/application/about.cfm?page=patients2

variations are determined through genotyping. Would it be really bad to explore, develop, and eventually practice individualized medicine? The science of genetics makes it possible soon—in 10 to 20 years—but a lot more than science is needed for individualized medicine to become a reality. Patients have to be convinced that they should permit genotyping. They have to be assured that neither the insurance companies nor their employer will use it against them. Society has to guarantee to protect the patient legally and robustly. A great benefit has to be demonstrated for the patient and the doctor and the insurer alike.

Now, in theory, once this has happened and individuals can be divided into small groups, the drug companies can make drugs for these groups of individuals, even if it might, at its extreme, be for a group of only one. The markets for these drugs will be smaller, but reduced sales will be compensated for by reduced marketing costs and, hopefully, improved market penetration. But the idea that drug companies should only make a drug that sells for a billion dollars has to be first forfeited by both the industry and their investors; there simply won't be any large enough indication encompassing millions of patients for the same drug once we are all genotyped.

Will drug companies like this? Not really! However, they will have another benefit: smaller trial sizes. At the extreme, in order to make the trial statistically valid, the trial size might include everybody who has the disease. This would appear to be a clear idiocy. But the idea is catching on in Europe at least, where the authorities will permit smaller trials with all treating physicians linked by e-mail. In this way, side effects can be reported directly rather than only through the European Medicines Evaluation Agency (EMEA). As a consequence, the few people who are on the drug can be immediately taken off, or checked for some crucial and predictive enzyme activity,[25] when any single patient reports side effects not yet known. A small, interlinked group of physicians and patients could respond very rapidly, preventing death or other severe consequences.

For individualized medicine to become a reality, all of these things—genotyping, assurance of patients' rights, research for small groups, and altered trial design and its acceptance by regulatory agencies—must happen simultaneously. Admittedly, it is about as simple as the Middle East peace process. Is society ready? Probably not yet.

[25] For example, glutathione-transferase induction in the liver.

3

HISTORY IS GOOD TO KNOW

CLASSICAL DRUG DISCOVERY

History is good to know; looking at history is rewarding. This is particularly true in the drug industry. Modern methods such as *high-throughput screening* (HTS), which involve mass automatic screening for candidate molecules, have great potential, which will be realized, but luck has featured very highly in the industry's successes in the past centuries.

A good starting point for history is the sixteenth century, when Paracelsus—most famous for his observation that toxicity is dose dependent—used mercury to treat syphilis and reintroduced opium to Europe.

The opium history is illustrative of progress and most of the drug discovery process. People were using opium—and its natural active ingredient morphine—for nearly 5,000 years without having any idea what was its key chemical ingredient or why it worked. It is very doubtful that modern medicinal chemists would have come up with morphine. It may also be unlikely that chemists can improve on morphine as a painkiller in the next 1,000 years. Its pain-killing effect is in fact a side effect of its first use in the Far East, namely, as an antidiarrheal agent. (See Box 3.1 on page 33)

It would take until the late twentieth century to understand the actions of morphine fully. The fact that morphine works is very informative. It tells scientists that there is a target for morphine and that this target

Table 3.1 Some Landmarks in Drug Discovery Pharmacology

Year	Landmark
1540	Paracelsus: used chemicals to treat disease (mercury for syphilis); reintroduced opium in Europe
1670	Wepfer: conducted systematic animal study of poisons
1803	Serturner: isolated morphine
1848	Merck: identified papaverin[26] from opium extract
1900	Schmiedeberg: studied the effects of chemicals on living organisms—giving birth to a new discipline, pharmacology
1910	Ehrlich: studied first antimicrobial agent, salvarsan, resulting from systematic synthesis efforts; gave rise to receptor concept and chemotherapy concept
1960s	β-blockers discovered: first "mechanism based, receptor subtype specific" drug class
1990	HIV protease inhibitors: "first designer drugs"; 3D-structure of biological targets known and used in drug design; recombinant protein drug: growth hormone (GH); therapeutic antibodies
2000–	Genomic-derived target and discovery-based drug used in the clinic and in diagnostics

would be a receptor for an endogenous substance. It wasn't until 1982 that the endogenous "ligand"—the opioid peptides *leu*-enkephalin and *met*-enkephalin—were found, and the receptors were cloned in 1992. A lot of drugs were used without knowledge of the biological basis for their actions. As long as safe and effective doses can be elucidated, clinical praxis proceeds. Physicians are practical people; they judge the therapeutic benefit, not the theory behind a drug.

Using drugs as antimicrobial agents started in 1910 when Nobel Laureate Paul Ehrlich and Sahachiro Hata found in a systematic search an arsenic-containing compound (Salvarsan) that would kill the syphilis spirochete bacterium but not the host. The science of medicinal chemistry—manipulating a chemical to improve its pharmacological properties by a synthetic organic chemistry route—had been born. Until this point, most experts thought that vaccination was the only way to protect against infectious disease. Since Langley's and Ehrlich's concept of the receptor was described, drugs were now being designed based on the structure of the target.

---

[26] Papaverin is a stomach and intestine relaxant—a phosphodiesterase (PDE) inhibitor.

Most of the 433 distinct chemical entities currently used as drugs were discovered using the classical drug discovery paradigm. They came from studies of toxins and herbs, and from the study of physiology and pathophysiology. Later, more drugs were developed by examining the side effects of existing drugs and then by studying the active metabolites of such drugs.

The basic idea is to find an herb or some other plant material or animal toxin that has a "pharmacological" effect, such as sedation, on an animal, and then isolate the active ingredient from the mixture present. One then determines the molecular structure and works out a way to synthesize it and finally demonstrate its biological activity. One can then try to modify the structure to improve it, that is, to make it work better, and at lower doses. This process works, just as it did for morphine for 5,000 years, without knowing what the target is in molecular or cellular terms. An inherent weakness in the methodology is that significant amounts of all the potential drugs need to be synthesized and one has to use a significant number of animals for testing. One wouldn't wish to develop many drugs by testing in whole animals. Modern methods reduce the need for animal testing in tune with the changes in society's attitudes.

Glorious Serendipity

The "glorious serendipity" by which many drugs, especially antibiotics and in particular penicillin, were discovered seems once again relevant because of recent interest in old antibiotics such as ciprofloxacin as a treatment for anthrax infection. Alexander Fleming discovered that penicillin is a "bacteriocide"; that is, it kills certain bacteria. Something that just stops bacteria's growth—a "bacteriostatic" agent—is not good enough. Today penicillin is still made the same way, namely, by fermentation. One single facility in the whole world makes most of the raw penicillin. Incidentally, it is located in Belgium in a former stable built by Napoleon for his horses and owned by the fermentation company Gist-Brocades. Also incidentally, but much more importantly, having few suppliers makes the whole world incredibly vulnerable.

The more recent history provides a lesson in how perceived and real medical need can drive production. In 1940, about 2,500 units of penicillin were made. But in 1943, the U.S. Army undertook impressive scaling-up and had made millions of doses per *day*. This was because they perceived that there was a crisis and that the Army was vulnerable. They could do this because it is easy to establish efficacy for antibiotics. Even then it took only three days. Today, it takes only six hours with a technique called *polymerase chain reaction* and known as PCR—the

Table 3.2 Antibiotic Activity

- Fleming discovers penicillin in 1928.
- Bacteriocidal activity is determined.
- Production of penicillin is still microbial by fermentation (Gist-Brocadees in Belgium is the only manufacturer). Military intervention to preserve sources in 1940 in Oxford, and in 1943 by the U.S. Army (which can be compared with the 2001 ciprofloxacin "crisis").
- Structure is determined and then can be synthesized.
- Efficacy is easy to establish.
- Biology of the bugs keeps the industry "alive."
- New antibiotics are produced to meet resistance development.
- Determining the site of action gives rise to new classes of antibiotics.
- Determining metabolism gives rise to new classes of antibiotics.
- Failure of new antibiotics in clinical trials is on safety or solubility, and virtually never on efficacy.
- Trials are short: 6–24 hours with PCR or 14 days with serology.

same technique that enables DNA profiles to be determined from microscopic samples—to show if new bacteria are growing. The other reason it could make this massive amount was because penicillin—even 60 years later—is made by fermentation much like beer. So not just beer is fermented in Belgium.

Unfortunately for society, the microbes develop resistance by changing their own molecular structure. But the development of resistance keeps the industry alive. If a pharmaceutical company develops antibiotics, it is assured of a continually developing market. The R&D team within the company must determine the site of resistance and then make a new class of drugs, which are easy to test for efficacy. Although it is easy to show efficacy—the killing of the bacteria—there are often technical problems related to solubility and administration, and there are serious problems of safety for new antibiotics. Not surprisingly, it is hard to kill an infection without having a toxic effect on the host. Many candidate antibiotic drugs fail on safety. However, given that it is seemingly easy and there is a need for continuous development of antibiotics, it may be surprising that worldwide only three to four large pharmaceutical companies make antibiotics. This makes the world very vulnerable, and regrettably, it makes the impoverished Third World even more impoverished. The fact that infections are much more serious in regions without adequate sanitation or water may be one reason why so few companies have an antimicrobial program for tropical parasites and microbes.

The Goodness of Side Effects

The discovery and subsequent development of antipsychotics is probably the greatest example of how drugs have been discovered through side effects.

After World War II in southern France, psychiatric hospitals were full, and people with hallucinations were often restrained. Patients developed rashes from the cloth that restrained them. The allegorical tale is that a kind doctor attempted to make them more comfortable with a newly discovered, newly synthesized compound—chlorpromazine. Chlorpromazine had been synthesized and described at Rhône-Poulenc reputedly by chemist Paul Charpentier and pharmacologist Simone Couvoisier and their colleagues, and was designed to be an antihistamine to reduce rashes and serve as an analgesic to help reduce pain. Indeed, chlorpromazine was also known to prolong the analgesic effects of propofol and was used in general anesthesia during surgery. Thus, at the famous French Military Hospital, Val de Grace, on January 19, 1952, Henri Laborit and his colleagues[27] gave this antihistamine to people who had "florid" hallucinations. Hallucinations are the most important "positive symptom" of schizophrenia—a common disease with a prevalence of around 2% of all populations in the world. The patients were all hallucinating vividly and unmistakably. While their rashes from the restraining clothes they were obliged to wear did not improve, their hallucinations were ameliorated by the chlorpromazine.

This *bona fide* antipsychotic property of chlorpromazine was established especially by two German studies, which looked at the dose dependence of the effect. Clinical use of this antipsychotic property of chlorpromazine spread between 1953 and 1955 through Europe to America via Canada where it was marketed by Smith Kline under the trade name Thorazine. Laborit can therefore claim to be the father of *psychopharmacology*, now a massive endeavor and industry, built upon the foundation of a poor antihistamine!

At this time, there was no knowledge of the molecular basis of chlorpromazine's effect or the molecular target. Knowledge of the molecular basis came 20 years later in 1972 when Paul Greengard's laboratory established the connection with the dopamine receptor D_1. This led the whole industry to look for a new antipsychotic better than chlorpromazine and to new agents tested directly in multiple clinical trials. Further research 20 years later showed that much of the antipsychotic effect was through another variant of the dopamine receptor, the D_2 receptor, not the originally

[27] Accounts vary in detail and in assigning credit. Some accounts credit the first use on a psychiatric patient to Hamon, Paraire, & Velluz. Further testing as an antipsychotic is credited to Delay, Denikes and Harl in a mental hospital setting.

proposed D_1 receptor. Connecting antipsychotic action with antidopaminergic action defined antipsychotic research for many decades to come.

The good fortune of finding chlorpromazine accelerated the development of new treatments. Without it, the link between the dopaminergic system and antipsychotic activity would, however, have been found by other careful observation.

It is indeed fortunate that progress of this type is not entirely dependent on unique serendipity. The connection between psychosis and dopamine would have been found, for example, because dopaminergic drugs used to treat Parkinson disease can precipitate psychotic effects, and parkinsonian-like symptoms can come from antipsychotic drugs. In addition, amphetamines and lysergic acid diethylamide (LSD) cause psychotic symptoms, and sooner or later we would have found that amphetamine's site of action was blocking the reuptake of monoamines, in particular dopamine, so that the physiological effect of dopamine was enhanced and exaggerated.

Incidentally, but importantly, we may have unlocked the wards of psychiatric hospitals, but we are still looking for the etiology of schizophrenia. That dopaminergic intervention is effective on florid, active, hallucinatory symptoms in 60 to 70% of patients indicates dopaminergic involvement but does not prove causality. But patients and physicians buy a clinical effect, not knowledge of the mechanism. Thus, antidopaminergic antipsychotics continue to dominate pharmacological treatment of schizophrenia. The treatment is now so widespread that it is almost impossible to find an unmedicated schizophrenic patient in the Western world. This treatment is also so expensive that in 2003 antipsychotic drugs provided by the state represented the largest single expenditure in several U.S. states like Kentucky and Nebraska.

A drug's unexpected side effect becomes its primary effect. A drug may have been approved for one condition but after approval may be prescribed for off-label use. Other drugs find a use after proving ineffective during a trial. This was true for sildenafil, better known as Viagra. It was being tested for cardiovascular disease, and since results were disappointing, the company stopped the trial and recalled the samples. Urban legend has it that male patients enrolled in the trial allegedly declined to return them. The reason why was established, and the rest is history.

The Drug's Fate Becomes the Drug

Today one of the safest and fastest ways of making a new drug is by finding and isolating the active *metabolite* or the active *enantiomer* of a drug mixture already known to work in the clinic.

Phenacetin (phenadine) was an incredibly efficient drug for the treatment of pain and headache. The structurally similar acetanilid was sold by all the companies that were in the business of selling dyes. They both worked fine on headache, particularly on migraine-type headache. They worked fine except they were so toxic to the kidney—"nephrotoxic"—that phenacetin was to be one of the first drugs to be withdrawn from the market. To produce a better headache pill, one would have to remove the kidney toxicity while retaining the analgesic effect. Julius Axelrod and Bernard B. "Steve" Brodie, working in the 1940s, looked closely at the metabolites of phenacetin, and they found that the metabolite acetaminophen is much less nephrotoxic than phenacetin, but equally effective on headache, and unpatented Tylenol was born. It was "unpatented" because they forgot or omitted to patent it. The times were different.

Somewhat related to this path of discovery is the path whereby a drug may comprise a mixture of two components which are the same in chemical composition but which are mirror images of themselves, so-called optical isomers. Often only one of the isomers is biologically active, and separating this *enantiomer* from the composite drug gives a new drug and, incidentally, extended patent life. The isolated active isomer can often be used at lower doses, and side effects are most likely lessened. With a mixture of the optical isomers, the body has to metabolize and eliminate both the active and the inactive components. The inactive isomer is likely to have *some* pharmacological effect beyond the intended effect and produce a side effect. The isolated active isomer is by this logic inherently better. The FDA is increasingly keen to approve the active enantiomer rather than the mixture. Several small companies make a living by isolating the active form of a racemic-mixture type drug. Sepracore is one company that started out solely by separating and synthesizing "new" drugs from racemic mixtures.

Pathophysiology Reveals Drug Targets

A further method employed successfully in classical drug discovery comes from looking at the pathophysiology of a disease.

In Parkinson disease, Oleg Hornykiewitz examined autopsy material from 180 patients and found a huge deficit of dopamine in the basal ganglia.[28] Existing therapy, which is a successful symptom treatment, is to replace this deficit by giving a *precursor* to the dopamine, L-DOPA. The L-DOPA is converted by the brain into dopamine. Alternatively, one can make artificial "dopamines," that is, dopamine receptor *agonists*, which have a very similar effect on the receptor as natural dopamine.

[28] Area in the brainstem involved in movement initiation.

These therapeutic approaches were suggested from the original work by Arvid Carlsson. Both approaches provide some degree of relief from tremors and rigidity for 15 to 20 years. Unfortunately, they do not stop or slow the disease, but as symptom treatments they are invaluable.

Autopsy studies on Alzheimer-afflicted brains some 40 years later showed a similar huge drop in neurotransmitter levels, but this time it was acetylcholine (ACh) in the *hippocampus*—which is associated with memory. Choline-esterase inhibitors, which stop the breakdown of ACh, were introduced and are presently the only symptom treatment available (Aricept, Excelon, Reminyl, Cognex). Recently, in 2004, memantine was approved as the first treatment to slow progression of Alzheimer's in mild to moderate cases. Work is ongoing to define acetylcholine-like muscarinic M-1 agonists and M-2 agonists that act on acetylcholine receptors directly.

Both of these examples of a drug discovery paradigm emerged from pathophysiological studies. In both cases, the therapies did not require the modern molecular biological techniques to clone the receptor(s), or need a library of a million chemical compounds. However, these findings led only to symptom treatments. They do not help us to design a cure because they don't tell us how or why dopaminergic neurons die in Parkinson disease or cholinergic neurons die in Alzheimer disease. The therapies don't even help to slow disease progression.

Virtues and Problems of the Classical Pharmacology Paradigm in Drug Discovery

Classical pharmacology has achieved remarkable successes. When you test a drug in a whole animal and look for an effect, there is an immediate emphasis on efficacy in reaching the target organ and determining acute toxicity. These important attributes are all determined at the outset or quickly afterward. This method requires an animal model of the disease in question. But it has inherent shortcomings. It worked at a time when no one was thinking of testing more than 100 to 1,000 compounds in a drug development program. The compounds tested closely resembled the original active substance, which is why so few chemical structures are in clinical use today. The compounds have to be synthesized in large quantities—many grams—in order to test each in whole animals or organ bath experiments. It regards a whole organism as a "black box." This is not satisfactory intellectually. The approach emphasizes the efficacy only of the compounds you give. You only get results from those compounds that actually do something in a whole animal. You might be trying to sedate, make it throw up, make it stop throwing up, and so on. This isn't always easy. The last example is very difficult to achieve in animals, yet it is very important especially as cancer chemotherapy is associated with severe nausea. Society needs a good antiemetic as an adjunct to cytotoxic oncology drugs, but it is very hard to develop. The best animal model for an antiemetic is a ferret, which naturally throws up in response to noise as part of its defense mechanism. Incidentally, the incorporation of guinea pig into English and other languages derives at least partially from its use as an animal model in biological experiments.[29]

To gain deeper insight, one needs to shed light in the black box. Why is this important? If a drug company didn't succeed with this blind approach, if it found bad side effects, it usually abandoned the entire program and tried something completely different. It wasn't, at this time, trying to produce too many drugs. This represents a great loss because some of the compounds made might have been structurally almost the perfect molecule to achieve the desired effect had they reached the target organ, but the blind black box experiment would not have shown or revealed this.

Merger Frenzy

Merger frenzy traces the economical background for the pharmaceutical industry in general. Drug discovery became a large industry some 120 years ago when chemical companies that were making rather cheap dyes turned to make more sophisticated products (i.e., added greater value). The result is music to any investor's ears. There is a continued willingness to invest in the capital-intensive Pharma industry. The medicine market has increased continuously in the past 50 years, doubling every 6 years. Other industries are inherently more limited in growth potential. Indeed, how many more cars can you sell? Admittedly, the costs of drug discovery have continuously increased, but the companies have grown to distribute the risk and to amass the necessary capital. Initially they could do this through organic growth, and then they turned to merging and making acquisitions. Growth, merger, and acquisition have gone in parallel over the past 10 to 15 years. Glaxo became

[29] The guinea pig is still the only useful model for so-called new NK-1 receptor-type antidepressants, since they have the same receptors as humans and the vocalization of pups separated from their mothers is the manifestation of "depression."

Glaxo-Wellcome, and SmithKline became SmithKline Beecham, and they both became GlaxoSmithKline (GSK). Sandoz and Ciba-Geigy made Novartis, Astra and Zeneca made AstraZeneca, and Rhone-Poulenc-Rohrer and Hoechst made Aventis, which was acquired by Sanofi-Synthélabo. Pfizer bought Warner-Lambert (Lipitor) and Pharmacia UpJohn (Celebrex)—and thus became the largest company by 2002—and so on and so forth. It's almost not worth remembering them because tomorrow it will change. National boundaries are mostly forgotten; they are true multinationals.

The consequences of this merger activity are not uniformly good. The stock market used to greet mergers with applause; now it is not too sure.[30]

The problem is that they have more money and they are bigger, but bigger doesn't mean they are better. How can they even keep track of what they know? This applies to the pharmaceutical industry fully. Once a company becomes so big with 70 to 100,000 employees and has so many research sites, there is so much knowledge, in so many places, that never gets to be used fully in the right place. Information management becomes a serious issue. The real evaluation of whether these mergers were good or bad for making new medicine has not yet been completed.

Although the quality of drugs has improved tremendously, the expectations of the stock market, patients, doctors, and government have grown even faster, which has created a gap. New technology-based drug discovery has permitted the birth of the biotech industry both to develop *biologicals* and to provide "platform" technologies for Big Pharma while leaving capital-intensive stages of drug discovery such as clinical development with Big Pharma. More recently, small companies have tended to want to do everything, including the clinical trials but excluding marketing.

And there are opportunities for Biotech. The consolidation within the industry has reduced the number of companies investigating particular diseases or conditions. For example, in cardiovascular disease there are now only three to four truly major players: Merck, Astra-Zeneca, Pfizer, and Novartis. In 1991 there were seven companies making antibiotics; now there are three, dominated by Pfizer and Abbott. The R&D programs of the acquired can be dropped. Pharmacia "merged" with UpJohn-Searle mostly for its COX-2 inhibitor (Celebrex); it dropped much of the rest of the research programs. Pharmacia is now part of Pfizer, and much of Pharmacia's research portfolio has been similarly dropped.

[30] For example, in the second quarter of 2003, Pfizer lost $3.67 billion, partly as a result of a drug that was a subsidiary of a subsidiary of a subsidiary acquired several mergers ago.

Box 3.1 Opiate Pharmacology in Reverse

2002	Although 4,000 ligands are known, we are still waiting for a good specific, approvable opiate receptor ligand for treatment of pain.
1992–1997	Opiate receptor subtypes μ, κ, σ, δ determined.
1992	First opiate receptor cloned.
1978	Other endogenous ligands: β-endorphins discovered.
1976	Morphine antagonist naloxone introduced. Morphine receptor subtypes suggested.
1975	Endogenous morphines, leu- and met-enkephalin, isolated using organ-bath assay.
1960–1970	Morphine receptor organ bath assay developed.
1910–1960	Biological effects of morphine cataloged on GI, pain, CNS.
3000 B.C.	Opium and its extract morphine known to be an antidiarrhetic and analgesic extracted from poppy straw.

The events in the story of morphine are in the reverse order of how we try to perform classical pharmacology, where we go from discovery and elucidation of the science to development of the drug. The morphine story starts 5,000 years ago with the poppy flower seedpod, which could be chewed or cooked to produce opium. Opium was recognized for its antidiarrhetic and analgesic effects. Morphine was then extracted, its structure was determined, and its effects were catalogued using an organ-bath assay and also through continued abuse. Morphine was at this time the most misused psychopharmacological drug by doctors and pharmacists and was coined morphinism. Endogenous morphines, the enkephalins, were found by Hans Kosterlitz and John Hughes, and independently, by Sol Snyder and Lars Terenius. A little bit later, naloxone was introduced to treat morphine addiction. We cloned all the receptors we could find and developed more agonists. We have about 4,000 ligands, but not one of them is approvable—*or at least not without serious lengthy warning labels about addictive properties and difficult prescribing procedures*—and **none of them is better than morphine**! Why are they not better? Because the efficacy of morphine is so fantastic. A slight improvement in side effects is insufficient justification to approve something with decreased efficacy. New molecular recognition techniques that match the ligand to specific receptors—in particular the μ receptor—may eventually yield a worthwhile competitor to morphine as we start to sort out which "morphine receptor" needs to be targeted for analgesia, for sedation, and for its addictive effects.

chapter

4

THE BETTER BETA-BLOCKER BARRIER

THE LARGEST SELLING BETA-BLOCKER[31] NOW COSTS 15CTS/DAY

One of the great pharmacological therapies that saves millions of lives by lowering the incidence of heart attacks and stroke are the beta-blockers developed by Sir James Black, who partly for this work carried out at Imperial Chemical Industries (ICI) in the early 1960s received the Nobel Prize in Physiology and Medicine in 1988. He showed that overactivity of noradrenaline[32] at the β-adrenergic receptor—a so-called G-protein coupled receptor (GPCR) of which we will write a lot—is one of the causes of high blood pressure and that using beta-receptor antagonists—"beta-blockers"—reduces blood pressure. Today these are cheap, reliable drugs for our stressed and aging population, whether used alone or in combination with other drugs. They are, of course, not without side effects. If you asked most members of society if it would be good to have a better beta-blocker—that is, a better drug for reducing high blood pressure—they would most likely reply: "Yes, of course, one with fewer side effects."

---

[31] More correctly, a *β-adrenergic blocker*. See indications, side effects, and adverse reactions at, for example, http://www.rxlist.com/bblock.htm

[32] Known in the United States as norepinephrine.

Unfortunately, this is becoming more and more unrealistic. The two negative factors are time and money, though perhaps not in the simple, conventional sense. "Not enough time, costs too much money" might be replaced by "Too much time, not enough revenue." Suppose you have a candidate drug, a chemical compound that could be shown in the laboratory to work better than the largest selling, in terms of tablets sold, beta-blocker, propranolol. How many people would you need to recruit in a trial to test your new improved cardiovascular drug to show that it is more effective than the beta-blockers and, in addition, the other major group of antihypertensive drugs, the angiotensin converting enzyme (ACE) inhibitors?[33]

For a new beta-blocker, you would need 40,000 people in a two-year trial, to be extended for eight years, to examine whether your new candidate drug changes the occurrence of cardiovascular incidents like heart attack and stroke. This is the equivalent of a **400,000-man-year** trial. This would be a huge trial. If this is not enough to dissuade a pharmaceutical company, consider also that these four successive groups of 40,000 must all not be taking beta-blockers or must be taken off them.

If you are also planning to prove yourself better than the competition, you have to test your drug not only against no treatment or a placebo, but also against the existing drug. If that drug happens to be expensive, then you will have to buy it from the competition. **There has to be something significantly wrong with the existing medication if you are going to successfully launch a competitor**. The result of all this is that beta-blockers and ACE inhibitors cheaply and safely corner the high market demand for control of blood pressure. The 40 million "patients" in the United States will not be offered anything new for probably quite a while.

Moreover, propranolol also has a number of *useful* side effects. It has been proven to be effective in **migraine** prevention, and any new drug for migraine prevention will also have to be better and cheaper than propranolol. Notwithstanding the success of the *triptans* introduced several years ago that work once the migraine attack has started, preventative migraine medication will not be found unless, of course, it is found to be the *side effect of a new drug approved and made available for another indication.*

---

[33] The ACE inhibitors very development at Merck showed that early on, when propranolol had patent protection and was much more expensive, a company that missed making a beta-blocker could successfully come up with an alternative drug for treating high blood pressure. Twenty ACE inhibitors and angiotensin-1 (AT-1 or ATR) antagonists were developed. Now beta-blockers, ACE inhibitors, and angiotensin receptor antagonists are generic and cheap. In Europe, allegedly better beta-blockers are routinely used, such as Enconcor, Seloken/Spesicor, and Atenblock.

Currently, there are no, or very few, new *targets* to treat migraine; a breakthrough in basic and clinical research is required. Propranolol also combats **angina** pain and reduces the risk of a **heart attack** and, once again, it is very cheap.

Despite these enormous obstacles of huge trials and cheap competition, new drugs can emerge, and sometimes clever trial design can be the key here. The first example is of a toxin becoming a "blockbuster" drug. Usually, toxins are only used in the laboratory to define a drug target, but in this case the toxin itself is used as the drug.

Ironing out the Wrinkles of Botulinum Toxin

Botulinum toxin (BT), "a great actor in history," is one of the key proposed reasons for population decline in Mesoamerica. BT is made naturally by the bacterium *Clostridium botulinum*; it is a very efficient and potentially deadly neurotoxin causing muscle paralysis. It still kills when food, especially canned food, becomes infected. It was first introduced for treatment of **pain** in 1983, but people didn't know how to dose it, and it had to be taken off the market. Rather creative people at Elan Pharmaceuticals realized it still had activity and found a small, rather uncommon—1 in 10–20,000—neurological indication for which it could be proved efficacious. They used it in the treatment of *cervical dystonia*,[34] and Elan had it approved by the FDA in 2001, having tested it in only 570 people. The patients needed only three injections, two days apart, into their neck muscles, an easily reached site. In addition, the patients assessed their own improvement. Such a self-reported effect is known as a *visual analogue scale of self-reporting*. It was a clever strategy. Once in the marketplace as Myobloc, BT was subsequently approved in 2002 as Botox for "treating" wrinkles. It is now the leading product of Allergan, a company specializing in ophthalmology and dermatology that has become very profitable. Wrinkles may not be as important an indication as pain, but very successful financially indeed. Incidentally, BT may have yet other uses in combating cognitive decline and schizophrenia.

[34] *Dystonia*, rudely translated, means badly functioning muscle tone as is seen in many movement disorders. *Cervical dystonia*, also known as spasmodic torticollis, is a focal—that is, it occurs in one place—dystonia characterized by neck muscles contracting involuntarily, causing abnormal movements and posture of the head and neck. (Adapted from http://www.dystonia-foundation.org/defined/cervical.asp)

Painful Processes

All clinical pain studies are using a visual-analogue scale. We are so incredibly bad at treating pain. The terribly sad situation with pain is illustrated by the following. To enroll in a study for a new pain drug, a patient has to have pain graded by himself or herself at level 4 or higher on a scale of 1 to 10, where zero is no pain and 10 is unbearable pain.

One of the most "efficacious" pain drugs, gabapentin (or Neurontin), was originally designed—and is still used—against complex seizures in epilepsy. It sells now for over $2 billion per year. When used, tested, and approved in a small trial for use in neuropathic pain, it reduced, on average, self-reported pain from 6.3 to 4.3. At a pain level of over 4, those who left the study, which led to the approval of the best pain medication, would still be able to enroll in a new pain study! At best, pain relief is relative. Pain trials have difficult endpoints, pain medicines are often overdosed, and pain patients are usually overmedicated and thus ineligible for the trial. Pain patients are almost never satisfied in their needs; they might never be satisfied.

Another outcome of the use of Neurontin in the treatment of pain—lower back pain—resulted in the largest fine levied by the FDA of $490 million in 2004 for promoting a drug for an indication for which it was not approved. Physicians are free to prescribe "off-label," but companies can only promote or market drugs for indications for which they have conducted successful clinical trials and won FDA approval.

There is more potential for ingenuity in trials, and one other example is in dental pain. This is how it works. The trial enrollees are given the new medication after, say, having a wisdom tooth extracted. The patient will self-dose it, that is, is told to take "as needed" but not to exceed four tablets per day. The patients will not overdose and will be happy when they can stop. You only have to monitor how many tablets they end up taking during the few days of the trial. You just count the number of tablets left in the bottle. It's a perfect "dream" trial for pain, if you can do it. Unfortunately, most things *do not* work on dental pain, but if they do, it is very effective: no hospital rooms and no expensive physicians required to count the remaining tablets. In many, many trials, of course, the physician is the one who monitors improvement using long-standing expertise and sophisticated expensive instrumentation. For example, imaging techniques such as magnetic resonance imaging (MRI) will cost the company conducting the trial $1,000 to $2,000 per scan per patient.

VERY PAINFUL PROCESSES

Everybody in the industry would like to make a drug that will cure a disease, or slow disease progression, or, at least, relieve the symptoms of the disease.

Effective drugs already satisfy many of the "simple" indications such as control of blood pressure, with β-blockers, angiotensin receptor (ATR) antagonists, and angiotensin converting enzyme (ACE) inhibitors, control of gastric acid secretion by proton-pump inhibitors—or "H^+-ATPase blockers—or with H_2-antihistamines, etc. Most of these drugs provide such a robust therapeutic effect that they can be tested and win approval as "stand-alone medications." That clinical practice often combines drugs that were approved as stand-alone medications has to do with many factors. Most importantly, sick people often have more than one disease.

Sometimes one may have a new opportunity to ameliorate disease with adjunct—or supplementary or additional—therapy. An example would be the use of a drug called carbidopa in Parkinson disease (PD). PD patients have reduced levels of dopamine in the brain, and treatment with levodopa (L-DOPA) increases the concentration of dopamine in the brain. Unfortunately, before it reaches the brain, some of the levodopa is converted into dopamine by the enzyme called COMT[35] in the periphery. Carbidopa is a COMT inhibitor that stops this breakdown of levodopa in the periphery so that more of the levodopa crosses the blood-brain barrier and enters the brain where it can be effective once converted by enzymes in the brain into dopamine.

This is good adjunct therapy. The drug (levodopa) is made more effective by a second drug (carbidopa), and the drug works at a lower dosage. But if you have to combine your drug with another drug, the FDA will not allow you to try your drug alone. As in the above example of trying to prove improved efficacy over a competitor, you will have to buy the drug from your competitor—unless, of course, it's one of your company's drugs—and supply it to your trial patients.

To make matters more difficult for "non–stand-alone drugs," an additional group of people has to be added to the trial. If you are trying to treat, say, rheumatoid arthritis (RA), a chronic progressive disease, the trial takes 24 to 48 months. All of the many patients have to be X-rayed, which in itself is not cheap, to see if the drug slows deterioration of their

[35] Catechol-O-methyltransferases.

joints. If you are competing against a newly approved and expensive drug, for example the tumor necrosis factor (TNF) soluble receptor Enbrel, then it will cost, per enrolled patient, $10,000 per year for the Enbrel. You have to buy the drug that you are trying to beat or equal. At the end of this very expensive trial you are expected to have shown that adding your drug is better than the using TNF soluble receptor alone. No wonder that virtually no one wants to develop an "add-on" drug. Drugs may end up as "add-ons" even when they were developed and approved as stand-alone drugs. All the trials have been paid for and they will bring in income. They are considered the "less successful children" of the company's drug program.

If you have had the courage to get this far and your new drug is approved, then you might hope that there will be a brave physician who will try your drug by itself and find that the patients improve with your drug alone. Alternatively, you could choose to start a new four-year study where your drug is tested alone. There are no guarantees of success at any stage. In some indications it is very tough to show improved efficacy and safety over existing robust, safe drugs.

Of course, it would be less expensive to develop add-on therapy to an older drug like methotrexate, which is comparatively very cheap. Incidentally, methotrexate is not even a fully safe drug. It was a drug developed for endocrinology and oncology that was found to work in treating rheumatoid arthritis. As such it is another example of "how drugs find targets" and a reminder that the industry has made so few drugs, only "433."

Exclusivity Isn't What It Used to Be

When Imperial Chemical Industries (ICI) first came out with a beta-blocker in England, the company benefited from it for 11 good years. It was the only antihypertensive; there was no other. Since then, the exclusive time in the market that any drug company has seen has been reduced enormously. This is because the Pharma industry has grown tremendously, while the number of targets to work on hasn't. When the new anti-inflammatory target Cycloxygenase 2 (COX-2) was discovered, the first of the new class of analgesics, the COX-2 inhibitors (Celebrex, Vioxx) enjoyed only three months of exclusivity.

For the companies coming into this new market space, it is of course important to follow quickly. Once the original drug's patent expires, then the generic competition can undermine any financial rewards. It is not surprising that companies turn to "chemical innovation" in drug development. It is a business formula borrowed from car companies. Chemical innovation is an elegant name for making use of a clinically validated target. The target has been validated by the other company's approved drug. Most helpfully, the other company's trial also shows you how to—or how not to—conduct your own clinical trial. One has to come up with a new, patentable chemical entity to the same [protein] target. Earlier, "chemical innovation" was called "patent busting." The affected refining of the phrase has not affected the process, which remains the same. As soon as the rumor reaches you about a great target, start to work on it and try to be first. If you missed being first and someone has already had a new innovative drug approved—one that will not become generic for a decade or more—try to come up with your own chemical version fast. The risk is smaller as the biology has already been proven. Patients benefit because the different chemical structure will carry somewhat different side effects while the main therapeutic effect is replicated. Thus, some patients may prefer the copycat drug based on their better tolerance of its different side effects compared to the original drug.

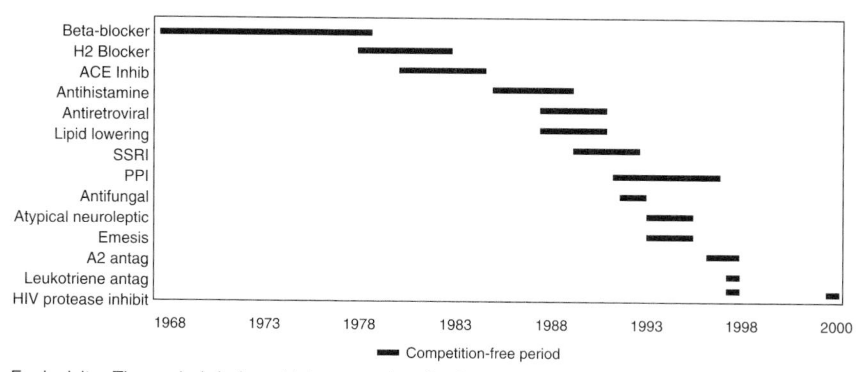

Exclusivity: The period during which a new drug is alone to address a given drug target. The competition-free period for β-blockers was effectively more than 10 years. Now successful drugs might be lucky to enjoy exclusivity for a year. HIV protease inhibitors, COX-2 inhibitors and Viagra are victims of their own success in more ways than one by encouraging competition development, which only lagged by three months for the COX-2 inhibitor.

Figure 4.1 Period of Exclusivity Is Shrinking

Unhealthy Competition

Companies keep a healthy eye on the competition. You may have run into research and development problems, had to modify your trial, or just been especially careful, but if you go to the FDA late and find that six other companies are before you, your drug will receive more scrutiny simply because the FDA and physicians learn about a class of drugs as they have been used in many patients. In principle, you only have to show safety and efficacy as the first drug in the class did, but, in practice, you will include a comparability study as you will wish to be judged to be better than the other class members. In addition, if your drug were the seventh *statin* for cholesterol lowering, you would really have to be much better than the rest to be able to market it well. There is a clear trend in the Pharma industry saying that "Best in Class" is better—for the bottom line—than "First in Class." Comparing some blockbuster drugs, the "best"—or most successful in terms of sales—is often the #2 or #3 in sequence of approval. It might be considered unfortunate that the innovative company doesn't profit as much as the "followers," but it might certainly be better for patients that the followers succeed in producing better molecules.

In Europe, the governments are increasingly saying that in each class of drugs they will only approve two or three different chemical structures because the side effect profiles and patient responses may be different for those. Any other drugs in the same class will have to be enormously much better. It is not yet quite at this point in the United States. Such decisions increase the pressure on company research to come up with new and original treatments, and, hopefully, to address diseases that they previously deemed "too small," "too difficult," "not sufficiently profitable," or "not easy—or treacherous—to perform clinical trials."

So it is very important that you keep an eye on the competition. In every drug company there may be 1 person in "competitive intelligence" for about every 100 scientists in R&D. The scientists in your company may know from rumors and published literature which compounds competitors have in trial and what happens in these trials. So they will know, for example, if Bayer stopped a trial, if Roche started a trial, and if Glaxo delayed a trial even though they may not know why. But the clinical chemist in the hospital where the trials are carried out may be able to tell in detail what happened. One might be able to gather more from personal research contacts in the companies, CROs, and hospitals than can easily be admitted. Putting together small pieces of legally available information often goes a long way. The discussion to publish all clinical trial data—successful or not—has been picked up again with prestigious influential journals such as the *Journal of the American Medical Association (JAMA)* and the *New England Journal of Medicine (NEJM)* among others, demanding such a register or database. Merck and Glaxo SmithKline (GSK) are already agreeing to it while the industry's collective body PhRMA[36] might still be "holding out."

Information on failed trials will be reviewed with great seriousness and gravity if any of these trials are in the same area, for the same disease, using the same targets with similar compounds as your own company's initiatives. The risks are great. If something goes wrong with a trial, the company may be sued for the worth of the company, not the specific damages caused. Therefore, pharmaceutical companies are very sensitive to disturbing news about any class of drug on which they also work or which they already have on the market.

If you are conducting a clinical trial, and a physician in the trial reports that one of their patients is not doing well, and you had just heard that another big company had stopped a clinical trial of a compound on the same target, what would the company do? There is a great temptation to stop the trial. It's probably the safest thing to do. Now, two big companies have stopped their trials. In addition, these two companies may somehow own the intellectual property on the candidate molecules and the target. *The chances of there being any further trials on the same target will be very slim.* If it turns out that this is the only decent target to treat this disease, nobody will dare, ever dare, to touch this target. These patients will never be treated using this logical, scientifically sound target. Rumors are lethal to clinical trials, to approvals, and to the clinical success of drugs.

Hopefully, the clinical trials have not been stopped for trivial reasons. That is why one needs courageous people within a drug development team, as well as courageous and attentive doctors conducting the trial, so that despite the warnings the trial goes ahead. If trials are stopped for the wrong reason, potentially good drugs are lost to patients and companies, probably forever.

Are there good, avoidable reasons why trials fail? They often fail because of side effects, which is a solid, valid reason. But other drugs fail because of underdosing; that is, a higher dose may have shown efficacy. They also fail because people don't have enough patients enrolled in the trial to prove efficacy, or to save money they designed too short a trial. The significant benefits might have become obvious only after twice the time and money were spent.

---

[36] Pharmaceutical Research and Manufacturers of America (see www.phrma.org).

Extensive and extended trials are the norm. In order to prove efficacy in major depression, a company would need to plan for around nine Phase III trials. Even 17 years ago, Prozac had seven, and most of the other SSRI-type antidepressants had nine trials, before accumulating three positive trials and receiving drug approval. The recent obesity drugs[37] had many trials, each involving 12 to 18 months and 20,000 people. The FDA rightly requires that these drugs show that the initial weight loss can be maintained for meaningfully long periods at meaningful levels of weight reduction. Big Pharma has learned to spread the risks; Biotech is often less well prepared. It often only has one target to work on, it has no clinical experience or big bank account for long, large trials. They have to team up with Big Pharma to mitigate risk and obtain a high enough level of funding.

Some Diseases Just Cannot Be Treated Today

Many diseases are too complex to be overcome by today's single-target medicines. There are too many poorly described variables interacting with each other and changing over a long time. There are too many potential targets that are inaccessible or too risky, and the likelihood of severe side effects is too great. Even if you could develop a drug to affect each target in the right way, it would be impossible under current regulations to conduct a trial for such a "cocktail" of medicines. The trials would need too many patients and physicians.

If a company, such as a small Biotech, is working on treatment of a rare disease, its competitors may simply enlist in its forthcoming trial everybody who has the disease. There is nothing the small Biotech can do.

Occasionally, opportunities arise because diseases are often not homogeneous. There are differences in the population suffering from the disease that can make part of the population treatable. Such stratification of patients can happen because a new diagnostic tool has become available that distinguishes these subgroups of patients. Having a more accurate selection and differentiation of patients actually improves the chances of a successful trial, so it works both ways: it makes the selection of patients harder, but the likelihood of success higher. Most Pharma companies prefer stratification of patients even at significant cost;

[37] Such as sibutramine, a.k.a. reductil, Ectiva, Reduxade, Meridia, and so on, and orlistat Xenical.

failed trials are so much more costly. Sometimes the discovery of a reliable diagnostic tool that FDA and the companies agree on unleashes a wave of research and trials. For example, the use of successive bone density scans as a measure of efficacy of osteoporosis drugs—rather than counting hip fractures over a much longer time period—is a major reason why osteoporosis is now being targeted by many more companies. The increasing need for an osteoporosis drug with the increased average age of the population was obvious to everyone for a long time.

Safe Drugs Are Tried for Many Underserved Indications

Many drugs are used for alternative indications. This always seems like good news for society but sometimes, at least, may be mixed news for the industry. Though immediately beneficial, it points to one of the reasons it is so difficult to make drugs against disease-specific targets: many diseases share partly the same targets.

In this way it is not very surprising that most antiepileptics are also used in pain treatment, that several *tricyclic* antidepressants are used in pain management, and even that some antiepileptics became mood-stabilizing drugs for bipolar disorder.

The *ion channels*[38] that these drugs affect are very broadly distributed, and there is a great opportunity for treatment potential in many indications if you target ion channels, but there is also a great risk for severe side effects. The discovery that the antiepileptic that a company is producing has an analgesic effect can be quite simply discovered and confirmed. The discovery can be made in the company's laboratories where a scientist tests the existing drug in several pain models. The confirmation of the analgesic effect can be made by surveying the patients already taking the drug. You can reach these through the neurologists prescribing the drug and ask: "Did you ever experience pain relief on this drug?" Some of them will answer: "My visits to the dentist do not hurt as much since I have been taking your drug." It would be so hard to get such a Proof of Principle in an independent study. It may not be a great scientific discovery, but it is terribly interesting! This *meta-analysis* of data shows that we often know more about drugs and patients than we consciously use. Many pain researchers have done meta-analysis of the effect on pain states of drugs being used for other

[38] Ion channels represent one of the common protein targets.

diseases in their tireless efforts to improve the quality of life of their patients. In a new clinical trial announced in August 2003 in the United Kingdom the MRC is now testing cannabis on postoperative pain to test the anecdotal evidence that cannabis has a positive effect on pain with fewer side effects, such as nausea.

There is more the drug company can do. A real example is now in clinical trial in which the company developed an antiepileptic that acts centrally on the brain. However, in its original form the drug itself had a sedating effect in higher doses. As a painkiller, you would want it to act peripherally and for a shorter duration. The chemists changed the structure by adding two *polar* groups (i.e., make it charged) on the molecule, and it now becomes peripherally active; the polar groups make it *lipophobic* (i.e., it doesn't dissolve in fat) and, as a direct consequence, it doesn't enter the brain.[39] This is the great innovation. The biologists test it for its affinity against the target and demonstrate its improved affinity, which means it will act at a lower dose. In order to save time and resources, the company *outsources*[40] a study to test it against 63 other ion channels. This is for the selectivity data the FDA and others will need. To test it as an analgesic, the clinical trial is designed to show an effect on postoperative pain, which is only a four- to seven-day event, so that the trial can be short. If it works on postoperative pain, which it did, one can then test it in clinical trials and seek approval for treatment against the broader indication of *neuropathic* pain.

This is a good example of applied thinking to maximize the utility of the products a company already has. Even biologicals may find new uses. This at first sounds surprising, as these proteins, hormones, and antibodies are more selective than small molecule drugs, (i.e., drugs that are low molecular weight chemicals). Nevertheless, one can make use of the multiple effects of hormones under different distinct pathological conditions. Growth hormone sells more for accelerated recovery from broken hip surgery than for dwarfism for which it was developed. Further to this, Lilly lobbied for and, somewhat controversially, secured its use on normally short children rather than just for children considered abnormally short. In effect, Lilly is alleged to be "inventing" or encouraging the definition of shortness of stature—not just dwarfism—as a "disease."

---

[39] To put it even more simply, drugs can get into the brain if they dissolve in the fats in cell membranes. Charged molecules are usually water-soluble (i.e., are *hydrophilic*) and tend not to dissolve in fat (i.e., are *lipophobic*). Adding charge—making the molecule an *ion* changes how the molecule is transported in the body.

[40] Big Pharma recognizes the value of outsourcing projects such as this; one cannot do everything "in house." Biotech companies rarely remember not to stretch their resources in this way.

Estimated Value of Medicine for Society

The drug industry needs to sell drugs. They are now creating demand by advertising on television and in weekly magazines aimed at the general population. If you now open a weekly in America, currently around six pages are devoted to drugs. It wasn't so 5 to 10 years ago. This is not something the companies like because this is a huge cost, often comparable or greater than the cost of research. But in a competitive market, they have to do it because the other guy does it. It's the same principle as operated for nuclear deterrents: "if the other guy has it, I have to have it." Often, it is not the most interesting or life-saving drugs that are having marketing dollars spent on them; rather, it is the antihistamines against heartburn (H_2), or against allergies (H_1), and cholesterol-lowering drugs, with rheumatoid arthritis (RA) drugs following closely behind. There might be some six or more advertisements each month for *Allegra* and six for other antihistamines. It is competition and the importance of the product for the bottom line of the company, not the medical need, that sets advertising budgets whose magnitude is close to the total R&D budgets.

The cost of drugs is almost invariably held up as being too high. It is a brave or foolhardy person who says: "The U.S. should spend more, not less, on prescription drugs," which is what PhRMA's president Dr. Alan F. Holmer said in his address to the Commonwealth Club on April 24, 2002. Could he be right?

Although a high cholesterol level is not exactly a disease, the drugs—the statins—that treat it are a $6 billion per year industry. These cholesterol-lowering medicines, at a cost of less than $3 a day, can help patients avoid coronary bypass surgery that would cost about $75,000. Medicine that can stop osteoporosis weakening your bones, at a cost of about $730 a year, can prevent hip fractures, which cost an estimated $41,000 per patient. A medicine can prevent brain damage in stroke patients, saving an average of more than $4,000 per month per patient. Not all prescription drugs are bad for the total health care costs or the economy. In addition, great drugs also have patent life limitation. Beta-blockers went generic, and so, recently, did the first statin Zocor. While this is a big—but expected—blow for its maker Merck, European governments rapidly made Zocor an over-the-counter (OTC) medicine and encourage its use at the patients' own cost. They can do this as Zocor is a safe drug proven over decades.

Preventative medicine—beyond the vitamins and antioxidants—is potentially an exceedingly lucrative enterprise for the drug companies because of its potential importance for society. What would be the *pharmacoeconomic impact* of a proven cancer-preventing drug? How many billions of dollars would be spent? What and how long would it take to assess its efficacy and quantify its risks?

chapter

5

WHY SOME GOOD DRUGS DO NOT GET A CHANCE AND WHY OTHERS FAIL

MANY CAUSES, TOO MANY REASONS

The United States may be the only country outside Western Europe[41] where influenza is taken so seriously that each year—since it changes year to year—a vaccine is made against what is thought to be the most likely flu virus and people go to their doctor for a voluntary inoculation.

Making the influenza vaccine is a highly seasonal business, and when the virus strain of the year is identified one needs the production apparatus to produce 20 to 100 million doses of this relatively simple and cheap viral vaccine. When the strain is mistakenly identified as it was in the 2003–2004 season, there is no time to "regear." One has to hope for some cross-reactivity of the immune response to the vaccine with the real virus. Much depends on a good flu vaccine or drug today when the population's average age is high and heart and respiratory disease are prevalent. While influenza is serious in most patients, it can be lethal in those who are compromised by age and heart or respiratory disease. The flu is a big health concern and big business.

[41] Where vaccines are certainly given to the frail.

How much, then, would society value an effective flu medicine? The drug industry puts a lot of value—and money—in it. In fact, in 2001, two companies launched competing drugs; they are both wonderful antivirals. One is called Tamiflu[42] and the other Relenza.[43] They are two of the nine drugs approved in 2001, a year when the industry had spent $26.4 billion on R&D. So these two drugs represent a large portion of the total outcome of R&D spending. How many people would one think have taken them since they were launched? One might guess at least 1 million and probably more than 10 million. In fact, it's probably still less than 1 million. Are they good drugs? Yes; they're "perfect." They have efficacy, and they have safety. They are marketed by GlaxoSmithKline (GSK) and Roche, both of which really know what they are doing—except they both forgot one simple thing. Both drugs have to be taken within 6–12 hours of onset of the symptoms.[44] Unless you are a pharmacist, at work, there is no way you can get it quickly enough. There is now a clever drive to have it available in the heart wards[45] of hospitals, but this is not the general big market the companies foresaw.

Each company probably lost $2 billion in projected revenues in the first two years after launch. It does no good to blame Hong Kong for not sending out the virus! How did this go wrong? Both Tamiflu and Relenza are also now approved for the prevention of influenza in adults and adolescents older than 13 years. But, again, is attempting to prevent only very affluent schoolchildren from missing a school day an important enough use for the drug? It is not what the researchers or the companies had in mind.

The researchers at Roche-Gilead and GSK had made a drug that inhibited a crucial enzyme—neuraminidase—in the virus, which is a valid target. This was a significant finding. It is notoriously difficult to produce antiviral medicines because the virus hides in the host's cells, which gives it an immediate high level of protection.

What do these examples tell us? Was the release of these drugs a mistake? For the two companies, it presented a large financial burden. But it's not harmful to have these drugs available, even though their use will probably always be limited. However, if the drug company had stopped the program prior to the release, would that have been immediately recognized

---

[42] Oseltamivir phosphate; http://www.tamiflu.com/Roche also recommends Tamiflu for treating patients over one year old. See also www.fda.gov/cder/drug/infopage/tamiflu/default.htm

[43] Zanamivir; see www.gsk.com/products/relenza_us.htm

[44] However, the authors note that both GSK and Roche as well as the FDA indicate efficacy if taken within two days of onset of symptoms.

[45] A cardiovascular ward is a very important place to be concerned about influenza, because it is very bad for already seriously sick people to be infected by any incoming patient. Both drugs may be kept immediately at hand in the refrigerator outside the heart ward.

as being for a solid, valid reason? Yes. It isn't sufficient to have good, safe, and effective drugs; the industry is expected to have good, safe, effective, and highly profitable drugs. Of course, since this was written, the 2005 flu vaccine production debâcle and fear of a flu pandemic precipitated by bird flu, these flu drugs, especially Tamiflu, are being stockpiled by governments at a level of 150 million doses. Tamiflu has become a "billion dollar drug".

Imagine if an academic scientist, while working on a target for a drug company, discovered an antiviral agent. As a result, the scientists who discovered it would expect the drug company to continue with the project through clinical trials, and, unless something went wrong on safety, the drug would be approved. But a company may have very good, though nonscientific, reasons not to go ahead. In an actual real example, the scientists certainly did not appreciate this decision. They simply did not understand. But from a business point of view it may have been the right decision. With antivirals, a pragmatic reason for stopping a program might be that there were legitimate concerns that by the time this "perfect" drug came on the market, some three years later, the viruses would have been resistant to it. It might have turned out to be a very good and now scientifically sound reason not to go ahead. Of course, in this and other examples, the company decision can seem very arbitrary to the scientists both inside and outside of the company, and it is usually taken without public explanation. These are not democratic decisions, but ones made by profit-driven private companies that are accountable to their stockholders and to the tax authorities but not the researchers.

Patients Do Not Take the Drugs during the Trial

Prospective drugs can fail to survive the clinical trial process for seemingly mundane reasons. Perhaps the most difficult to guard against is "patient noncompliance": the patients do not take or continue to take the medicine. This can happen whenever the administration of the drug cannot be adequately monitored or supervised. It may be because the drug has already worked, and the patient saw no reason to continue, or that the side effects were too pronounced, especially if the dose was wrong, or even that the pill was unpleasant to take because of bad taste or because it was too large. Big tablets deter adults as often as they lead to syrup formulations for children. Unfortunately, patients are often not honest about their compliance and will claim to have followed the specified course of treatment. This is equally true for patients who fight hard to be included in clinical trials of a new experimental drug.

Whatever the reason, this makes it much more difficult to prove efficacy. Trials have to work statistically and if some of your sample is noncompliant, efficacy will be as hard to prove as if the medicinal chemists and "pharmacokineticists"[46] had calculated the wrong dose to give.

The other obligation for trials is that the drugs should be tested against placebo and the patient and physician should not know if they are taking the drug or placebo (i.e., the trials are "blind"). In the case where a patient detects side effects, this "blindness" disappears. Of course, patients taking the placebo often complain of side effects as well. Statistically, the case is harder to prove when the side effect is obvious, such as when the drug has a metallic taste and the placebo does not. One has to think of everything to make sure the placebo isn't detectable by other means. In terms of taste, strawberry flavoring is a popular disguise. If one hasn't thought of everything, the result might be that perfectly good drugs may fail in trial and programs that should have worked are dropped.

Patients Are Taking Other or Alternative Medicines

Another, more serious reason why drugs fail is because of interactions with other drugs, either other prescription medicines or over-the-counter drugs or even rather uncontrolled herbal remedies. At the trial stage, this might be controlled. Patients can be excluded or removed from a trial because they are known to be taking other medications. But after approval when the number of patients who can be prescribed a drug increases dramatically, the control a company has as to who takes the drug is removed, but its liability is not. Drug interactions are the single most common cause of withdrawal of approved, marketed, financially successful medicines. For example, on August 8, 2001, the FDA announced that Bayer had voluntarily withdrawn the statin Baycol (cerivastatin) "from the U.S. market because of reports of sometimes fatal rhabdomyolysis, a severe muscle adverse reaction from this cholesterol-lowering (lipid-lowering) product." The FDA went on to say that it "agrees with and supports this decision," explaining that "Baycol . . . initially approved in the U.S. in 1997, is a member of a class of cholesterol lowering drugs that are commonly referred to as 'statins.' Fatal rhabdomyolysis reports with Baycol have been reported most frequently when used at higher doses, when used in elderly patients,

[46] Literally, those who look at the drugs' speeds of distribution, binding, and actions. Every drug has associated kinetics.

and particularly, when used in combination with gemfibrozil (LOPID and generics), another lipid lowering drug. FDA has received reports of 31 U.S. deaths due to severe rhabdomyolysis associated with use of Baycol, 12 of which involved concomitant gemfibrozil use." They informed the public and its physicians that "there are five other statins available in the U.S. that may be considered as alternatives to Baycol. They are: lovastatin (Mevacor), pravastatin (Pravachol), simvastatin (Zocor), fluvastatin (Lescol), and atorvastatin (Lipitor)."[47]

Was this tragedy avoidable? Possibly. Is Bayer responsible for patients taking a higher than recommended dose, or taking it with other medication aimed at the same result? It is certain that Bayer did not intend to do this, but it is equally clear that Bayer might be, fairly or not, accused of "negligence." Bayer has already settled a class-action suit with 100,000 patients, but tens of thousands are suing the company individually for $1 million or more for allegedly having caused their disease or for having endangered them unnecessarily. Consequently, Bayer's main financial effort is to put funds aside to settle these cases. Insuring oneself against such suits is not possible; settling is likely to be cheaper than the premiums would need to be. The makers of Mevacor, Pravachol, Zocor, Lescol, and Lipitor will hope that their drugs will remain uncompromised and above suspicion. If the Baycol tragedy had happened before the development and release of these other statins, some of these other companies may have canceled their programs because, as we have mentioned before, failed drugs promote risk aversion throughout the industry.

At the core of this example is one of the underlying fundamental reasons as to why it is so important for the industry to discover new targets and new compounds to address, affect, and otherwise interact with these targets. There is much less chance for interactions with structurally dissimilar drugs acting on different targets. Currently, one liver enzyme, called cytochrome P450 (or, more specifically, CytP450 3A 4), metabolizes 75% of marketed drugs. This tells you that in certain respects the drugs must be similar since only this one of our many liver proteins "removes" these drugs from the circulation. This enzyme did not "evolve" to handle these particular "xenobiotics." The 150-year history of pharmacy is a ridiculously short time on an evolutionary scale. Cytochrome P450 is normally involved in the metabolism and modification of many important endogenous molecules, including, for example, steroid hormones. Thus, we do not want this enzyme fully "clogged up" with any of our drugs. Luckily, this enzyme is "inducible" in that the number of enzyme molecules can increase when there is more to

[47] *Source*: http://www.fda.gov/bbs/topics/ANSWERS/2001/ANS01095.html

do, as there is during drug treatment. Such increases need to be reversible upon cessation of drug treatment.

Drug interactions are dangerous because they can grossly distort the effective dose of the medicines and their resident time in the body.

Diagnosis before Treatment

One way the previous scenario might be avoided in the future is through better diagnostics. If all the patients entering a trial, or being prescribed an approved medicine, were known, with absolute surety, to have a particular indication and not to have a contraindication, then the results both of the trial and after approval would be likely to be much better. The FDA is also responsible for approving diagnostic tests. Some of these diagnostic tests may become very useful tools in stratifying or selecting patients who should and should not get the drug. As soon as a diagnostic criterion is approved, trials have to change. The FDA and the pharmaceutical industry usually agree on the improvement that a diagnostic test and measure provide for better, safer, and thus more successful, trials.

This is a step toward the concept of individualized and genomics-based medicine, and there are now a few examples where genomic diagnosis is obligatory before the start of treatment. We hope that there will be many more.

One of the first examples of genotyping as a prerequisite for getting treatment, and hopefully, eventually a cure, is in breast cancer treatment with Herceptin. Breast cancer patients are required to have been tested to show that they carry the targeted gene product (HER2) before they can be treated with Herceptin (trastuzumab) an anti-HER2 monoclonal antibody. There is no point in this expensive treatment regimen if the patient's breast cancer is not a result of this mutated gene. The result of using genotyping should be improved therapeutic efficacy in the treated group.

Interestingly, there is no link between the manufacturers of the test, Herceptest from Dako A/S in Copenhagen, Denmark, and the providers of the Herceptin treatment Genentech, Inc. in DNA, south San Francisco, California. This would seem to be wise, and indeed the FDA would not permit the same company to have a monopoly on both the test and the drug prescribed based on the test.

chapter

6

About the Economics of Target and Clinical Candidate Selection

Pharmacoeconomics 101: Some Underappreciated Truths

The same ingenuity and the same science go into developing treatments for small, rare indications, as go into big, common indications. The small indications will cost almost as much as large indications up to the stage of Phase II clinical trials. Developing drugs against relatively small indications such as epilepsy or stroke, with an expected market of $200 million per year, may not be substantially less than for a large indication such as obesity where the market may be projected to be $4 billion per year. Not surprisingly, Big Pharma is biased toward large indications because of the costly marketing machine. It already has 5,000 to 10,000 sales representatives for a large company, and it needs a higher profit potential.

In contrast, in life-threatening diseases, especially in the absence of other treatment alternatives, side effects are understandably more easily forgiven by the regulatory authorities. This is why small companies flock especially to oncology because they know the biology, they pioneered the new target, and because they expect "easier" FDA approval. Smaller and cheaper trials make it possible for a small company to go to Phase II and possibly Phase III independently of Big Pharma. Oncology trials might typically involve less than 100 patients. Small companies can afford to do it, the drug doesn't have to be "super-clean," i.e. devoid of any side effects, and the trial may last only three to six months. Remember that prolonging survival by three to six months is already an approvable criterion, and when the prolongation is one, or three, or five years, we all celebrate.

This is not a bad scenario; it is just business economics. There are selected opportunities for Big Pharma and plenty of opportunities for smaller companies. A significant drain on resources occurs whenever a drug fails after Phase I, but worse, (i.e., much more expensive) is to fail in Phase III. That is very expensive.

Why Do Drugs Fail after Phase I?

In indications such as diseases of the central nervous system (CNS), the respiratory system, or the cardiovascular system, failure is often through the emergence of side effects. This is because the intended targets have other biological functions in other organ systems. The side effects come directly from the biology of the systems, they are "mechanism-based."

Whereas Phase I trials may fail because the dose was pushed too high and the side effects are too great, many Phase II and Phase III trials fail because of underdosing. Dosing today is in some circumstances estimated or judged using modern methods. One can look at the binding of molecules in the actual patient using positron emission tomography (PET). With a labeled ligand under optimal conditions, one can visualize occupancy of the target site with the PET labeled drug. One can even make valid estimates of when side effects may be encountered with an increased dose.

There is a reason germane to this. Many "judgable"—or dynamically measurable—indications, such as blood pressure control, are already currently well satisfied with existing drugs. Other indications are much harder to judge or monitor. For example, congestive heart failure trials may require long, slow, dangerous clinical studies with high mortality. It would involve cardiovascular targets that are difficult to reach and modify. There is bound to be a higher risk of failure during Phase II. Phase III failures are even more expensive and regrettably common in central nervous system diseases, i.e., neurology and psychiatry, endocrinology and metabolism, and cardiovascular trials.

It may seem trivial, but they may fail simply because there are not enough patients in a pivotal trial. One always plans for a certain dropout rate of, say, 20%, but at the end of the trial, there may not be sufficient patients on the drug—or even alive—to show statistical significance. Only at the end of the trial do you know if you estimated correctly the number needed. If the disease you are trying to treat is a progressive illness, it becomes increasingly more difficult. In Parkinson disease, Alzheimer disease, or any chronic disease, the patients are aging as well as sick. To show a slowing of the progression of disease, you need a trial of at least three years, and you would be wise to project at least a 30% dropout rate if only because a number of patients may die during the trial from other causes.

If the indication is large enough, and one can afford additional trials, then one might overcome this, but, as the costs escalate, the potential return on investment reduces as the patent life decreases. As mentioned earlier, new drugs for depression are now expected to require five to seven large, expensive trials. For a good new indication on the existing label, however, one is willing to initiate a new trial, especially as that extends the "use patent." In a Phase IV trial for Prozac, Eli Lilly extended the label to cover several other diseases, some of them newly described—some allegedly newly invented—diseases such as *social phobia*. Similarly, Novartis has attracted some criticism for its marketing of Ritalin for attention deficit hyperactivity disorder (ADHD).

Official Causes for Attrition

More drug candidates fail inside the companies than outside. What are the official causes for attrition? The most common is from so-called absorption, distribution, metabolism, and excretion (ADME) data, which account for almost all antibiotic failures and about 40% of other drugs. About 25% fail through lack of efficacy and 20% on animal toxicity and preclinical safety. More than is acknowledged, failure is often from political reasons within a company. This may be purely for marketing considerations (a competitor has launched a better drug or another adequate drug has "gone generic," or gained extended labeling), or it may be because factions within a company want to put the company's efforts into another therapeutic area or another drug.

Lack of efficacy is not as solid a reason as might be surmised. If physicians conducting a trial test the drug in mild cases—that is, in patients who would recover on their own—how can the drug be shown to be better than the placebo? The drug will fail on lack of efficacy. If the dose given is too low, the drug will also fail on lack of efficacy. Unfortunately, the people really motivated to show efficacy, namely, the research and development team that did all the work necessary to put the drug into the trial in the first place, may have a very small say in matters at this point. Others involved, such as the Clinical Research Organizations (CROs) conducting the trial might be almost indifferent to such failures. As stated before, they are paid according to the number of patients entered into the trial and for conducting the agreed tests. The end result of all these factors and influences is that the program is stopped, most often irrevocably. It will not matter that the drug may actually have been efficacious if tested properly.

Testing in mild rather than severe cases presents a difficult dilemma for some conditions. Mild stroke may seem like an oxymoron, but in many cases people do recover with almost no detectable residual deficit or disability. Another is the bacterial infection sepsis, commonly known as blood poisoning, which in its severest form still causes many deaths, but in its mild form may not be especially dangerous in otherwise healthy patients. The alternative to testing in mild cases is to test in severe cases. Unfortunately, in these two examples, any trial restricted to severe cases will be measured against mortality, and no matter how carefully one writes the report, it is not a good outcome if your drug is used on people who eventually die, even if the number of fatalities has decreased.

The ostensibly business problem of where to allocate company resources when choosing between two programs may not be as benign as it sounds, and it can be resolved using tactics more normally associated with politics. The political solution is ancient in origin and could legitimately be called sabotage. Consider this scenario. The central nervous system team ("Team CNS") is anxiously working on its anxiolytic drug (i.e., one to reduce anxiety), and the cardiovascular disease team ("Team CVD") is calmly working on its antihypertensive. It is likely that the company will take only one of the drugs through full clinical development in a given year. Hypothetically, Team CVD can take Team CNS's drug and demonstrate that the candidate anxiolytic can invoke, say, arrhythmia in their test conditions. No matter that they used 500 times the intended dose for the anxiety treatment, the result is forever in the company files. If the anxiolytic goes on to cause arrhythmia in a patient during a trial, the company would be in trouble. The remotest possibility of this means that the drug is "dead." The official reason is "preclinical safety." But, in reality, Team CVD has successfully sabotaged Team CNS's drug.[48] "Failures" are all too often generated internally and rather ignobly.

[48] Team CNS might now need Team CVD's antihypertensive for self-medication.

Many retiring executives ask that failed drugs be outlicensed to them when they leave the company. They often ask for drugs they know failed as "victims" of sabotage or misjudgment, and not wholly surprisingly they start biotechs—based upon the resurrected drug candidate—which have become—in one example—worth 6 billion Swiss Francs.

Reasons for Stopping Clinical Development and the Fallout from Failure

Competition is a very solid business reason for stopping clinical development. Other competitive drugs may have been recently approved and marketed. Your drug may have insufficient qualities to distinguish itself from another new drug or from preexisting generic or nongeneric competition. There was a time when drug companies always wanted to be "first in class." This is no longer true especially since there are many cases where the second or third drug in a particular class makes more money than the first.

Adverse reactions are common at any stage of development and marketing, and this can cause cessation of a project. Prudent research teams have a so-called backup compound. Since the adverse reactions are common, companies insist on a backup even before considering taking a compound into clinical development. Many companies have spent years to discover a new target and went on to select just one clinical candidate that failed, say, for some nonmechanism-related toxicity. Other companies had found better compounds for their target. Even a small company should have a backup molecule. Regrettably, many Biotechs fail to do this perhaps because they have so much conviction about the success of their principal molecule or so little time and money to develop a backup.

The other type of adverse reaction comes from the company personnel themselves, and it might correctly be called a fear factor. They hear rumors of another company's failure and extrapolate that to their own; it fuels their own doubts. Companies failing with a particular target may taint the target forever. Good intelligence is essential: the other company might just have had a poor compound, or they in turn had become frightened for some unknown reason or a result of a similar rumor. They may have just changed direction and dropped the program. There are already too many legitimate chances to fail; irrational reasons should be avoided. History will not look kindly on each and every target that was dropped for the wrong reasons. If history decides that this was in fact the best target, a whole generation of patients will have been denied an opportunity for treatment. If another company has stopped the clinical development, we hope that the competing companies' first assumption is that it was because it was a bad molecule, not a bad target. Of course, if the first company holds a patent on a target, for which it only has a bad clinical candidate, it becomes an obstacle for all. In addition, companies, whether small or large, try to block targets from potential competitors through patenting tactics.[49]

One needs conviction to move forward in clinical development. It bears emphasizing that one needs compassionate, courageous, and scientifically well-trained clinicians to do your clinical trials, and the strength of their conviction will help in determining how to proceed. Even if you are working through a CRO it is good to have some direct access to the physicians. The better you know them, the better chance of success.

If you work in Big Pharma and have doubts about a candidate molecule, it is very difficult to distance yourself from it. You cannot even "safely" license out the compound to a third party because, in the end, people sue the company with the deepest pockets and no amount of protective paperwork will ultimately protect Big Pharma.

[49] For example, the corticotrophin releasing factor (CRF) receptor patent was held by Neurocrine, and development of candidate molecules by others was thereby effectively inhibited.

chapter

7

Target-Based Drug Discovery: Part I

Improving Therapeutic Ratios[50]

In preceding chapters we have mentioned that new drugs—New Chemical Entities (NCEs)—are very difficult to find and that drug discovery is the most regulated human activity. There is a connection. One reason that NCEs are so elusive is that the very efficacy and safety requirements dictated by regulatory authorities—and overshadowed by the fear of litigation—became, over the last few years, extremely difficult to fulfill. Today you can have a drug that works fine in 100,000 people, but if one person—because of some genetic disposition, some bad luck, or even some bad meal, which interacts with this drug's actions—has a very serious side effect, or even dies, the decision is that this person's severe reaction or death is so unacceptable to society that these other 99,999 patients have to come off the drug too. Of course, Big Pharma companies fight hard for drugs that are already on the market. Taking a drug making $1 billion or more per year off the market, after all the development and initial sales and marketing costs, represents a real and stunning loss. Companies will seriously question if

[50] While *therapeutic ratio* has a narrow definition in FDA regulations, simply put, and in this broad context, the therapeutic ratio is the therapeutic benefit over the side effects at a given dose of the drug.

the alleged side effects and/or deaths can indeed be related to their drug. But they also have to be very careful and will rapidly withdraw a drug if it becomes certain that there is a strong connection, or if they believe that is what experts of the many class-action suits against Big Pharma will claim in court.

It is not a question of whether this is right or wrong—Victor Hugo[51] eloquently enunciated the moral position with his "one child's life is not worth advancing the universe"—but once 100,000 plus people are denied a drug that does them good, taking them off the drug is not that "nice." There are all kinds of middle ways to go. But it is really a bigger and bigger problem because the regulatory authorities, and ever more so, trial lawyers, are increasing their demands and the requirements.

The other reason, also often mentioned in earlier chapters, is that there are really not so many clinically validated and "drugable" molecular targets. Despite all of the nearly 30 million or so compounds synthesized and recognized as chemical entities with pharmacological potential, medicinal and combinatorial chemists haven't really exploited the chemical universe in terms of molecular diversity. The extremely large number of compounds made does not indicate complete coverage of the chemical possibilities; there is still a lack of diversity in the collective chemical libraries of the pharmaceutical companies. There is still a valid expectation that new compounds—whether synthetic or natural products—can be found to act as good drugs at targets that the industry has so far been unsuccessful in hitting with safe and efficacious compounds. New chemical libraries routinely are screened against valued important targets. In addition, there is a new set of targets—those from the Human Genome Project and from the association of disease and genetic variation studies—that is enlarging the scope of interesting targets. The good news is that the industry hasn't yet made full use of genomics-derived targets. At the time of writing, actual trials using drugs aimed at targets discovered by the Human Genome Project are not yet underway on any large scale.

The industry is improving therapeutic ratios—the ratio of efficacy vs. side effects. It is easiest done with "biologicals": naturally occurring protein molecules (e.g., beta-interferon, erythropoietin [EPO]) that can start out as "better" drugs. We are not referring here to biological molecules such as steroids or natural hormones, but to proteins with very specific roles in human physiology and disease states.

---

[51] Victor Hugo (1802–1885) writing about the war of 1870.

The Relative Ease of Biologicals to Become Approved Drugs

The biologicals we are referencing are proteins of a size of some 30,000 to 180,000 daltons as opposed to the usual typically small drug molecules of size 300 to 400 daltons. They are endogenous—meaning they exist naturally in the body rather than xenobiotic meaning "foreign"—biologically active molecules that have been intentionally introduced as drugs. The advantage is that biologicals can be expected to have good efficacy and few side effects. But they have an inherent problem of being more difficult to administer. Proteins are digested by the gastrointestinal system; therefore, biologicals usually have to be injected. By themselves, biologicals are safe, efficacious, selective, and not too expensive to manufacture anymore. They are not totally risk free, however, since they have to be produced in absolute purity. If they are made cheaply by biotechnological means within bacteria, then they must be bacterial endotoxin free—meaning containing no naturally occurring toxin. Such toxins would cause fever and adverse immune responses such as inflammation. They must also contain no variant proteins that may interact with the human immune system in unpredictable ways. The last thing that would be wanted would be an adverse immune reaction to proteins the body normally should recognize as normal.

Biologicals have been approved for small, fragmented indications, yet still some have earned over $2 billion per annum and have established significant Biotechs such as Genentech, with growth hormone and tissue plasminogen activator (tPA), Amgen, with erythropoietin (EPO), and Biogen, with *beta*-interferon. Genentech realized from the clinical literature that tPA is good in dissolving blood clots and made it by recombinant means for treatment of ischemic heart attacks. It is also now used in very selected cases of stroke. It can only be used for ischemic strokes caused by blood clots blocking blood flow to the brain. While this may be the more common of the types of stroke, the fact that tPA must not be used in all types and must be used within three hours of the onset of stroke means that it could only effectively be used in c. 2–8% of all strokes. When it works, it works unbelievably well. For these particular stroke victims who are fortunate to be treated quickly in areas where there are good stroke units employing these treatments, it is good news indeed. From a business point of view it is a great example of line extension, that is, when a safe and useful drug, whether a biological or chemical, is tried in new indications and finds new patients, new diseases, new markets, and generates new income without additional expenditure on the basic research side. Of course, there are the associated development costs of new clinical trials for the extended label for the new indication.

So-called recombinant technology has advanced so much that in its eagerness to embrace the technology and produce new biologicals (which, incidentally, seem to pass FDA with much greater ease than new chemical entities), the industry has built more capacity to make extraordinarily highly pure proteins than anybody wants to use. Whenever a new set of biologicals is developed and approved, the capacity does come under some initial strain and, suddenly, is limited for a while. This is because producing proteins by recombinant means (i.e., by fermentation) is a capital-intensive business requiring almost as expensive fermenters as a good brewery. Indeed, both industries often use yeast to produce their best products.

Better to Let the Market Decide?

Therapeutic ratios are incredibly difficult to determine in clinical trials. This is partly a statistical problem and partly a limitation of practical trials. Practical therapeutic ratios for particular patients are determined when drugs are in use, that is, when they are being prescribed by practitioners. Physicians can adjust the dose to achieve the best therapeutic ratio for the given patient, taking into account the weight, age, disease state, and other medication being deployed to treat the patient. This is why the FDA and the pharmaceutical companies only recommend doses. Good physicians caring for patients can evaluate real efficacy. Of course, it is worth remembering that they cannot evaluate the benefit if the drug is not available. This brings us back to how difficult it is to have drugs approved and made available.

Is the pharmaceutical industry the only significant industry where "the market doesn't decide"? For example, no airline or airliner safety and use is similarly regulated. The pharmaceutical industry is almost obliged by regulations to produce a drug equal to or better than all existing treatments. On the surface, this may seem reasonable and "protect" society, but it puts extra constraints and restrictions on drug development and prevents "free" competition on price in most cases. The constraints themselves will reduce the number of drugs available.

There are at least two prominent reasons why society should be concerned about this. Whenever the market—that is, the combination of physician and patient—has been given a choice, it has chosen the best drug available. The best example of this, which we will discuss more fully later, is that of omeprazole (Losec, sold as Prilosec), which entered the market of gastrointestinal medicines against hugely successful, safe, and extremely well-marketed drugs for treatment of heartburn (Zantac and Tagamet). Makers of these drugs, which have a different mechanism of action, have spent an estimated $1 billion to countermarket the newcomer, in a rather negative campaign. Nevertheless, Astra-Zeneca's Prilosec and its "me-too" equivalent proton-pump inhibitors (Nexium, Prevacid, Protonix, and Aciphex) generated a market of $13 billion per year (in 2002) because the market decided they were the better than cheaper drugs aimed at the same indication but with different modes of action.[52] Indeed, as a result of the entry of Prilosec and other proton-pump inhibitors into the market for treating heartburn[53] and ulcers, medical practice has radically changed and the number of gastric surgeries for bleeding ulcers fell so rapidly that today few M.D.s who were trained in the 1990s or later have seen gastric surgery, which earlier was the most common use of surgical theaters. Being able to avoid the risks of all surgeries by "popping a pill" permitted a relatively expensive pricing for Prilosec, and still proved economically beneficial for society. *Prilosec is the single most important reason for the overcapacity of surgical theaters built before 1980.*

The other reason is that society has benefited from more choice whenever established drugs find new indications. Several examples have been mentioned already. Another would be the so-called tricyclic antidepressants which have found and developed a bigger market outside of their intended indication in pain treatment. It is no exaggeration to say that eventually every safe drug gets tried in pain or in other underserved indications through physicians' willingness to try it.

Of course, the new generation of antidepressants (the SSRIs and SNRIs, e.g., Prozac), which have replaced the tricyclics, came into being because the industry recognized the continued need for better molecules in conditions where not all patients responded adequately to existing medications. This is a market-driven phenomenon. The question remains as to whether the pharmaceutical companies should be obliged to jump through all the hurdles required to demonstrate that the drug they are developing for a condition is better than the competition. The FDA checks safety and efficacy; the company marketers want to be able to compare and positively discriminate their drug as being better, whether just by being more convenient or by being fundamentally more effective or safer.

---

[52] So-called histamine 2 (H_2) antagonists. Nexium is actually an enantiomer of Losec.

[53] Also known as "acid reflux disease" as advertised, at the time of writing, for example, by the makers of Prevacid in TV commercials in the United States.

A positive distinction is especially required if some of the competitors—often in the same class of drugs—for an indication are generic and significantly cheaper.

In a risk-averse industry, it becomes more likely that potentially useful and safe drugs are being shelved well before their utility can be demonstrated. It's always easier for a company to fail a compound than to let it go into the market and "keep an eye on it." There are no perfect compounds, and *every sale and use of a medicine is both a profit-generating and liability-generating step*. You have to hope that the next generation of medicine will be better and will have a better therapeutic ratio.

chapter

8

CHANGES NEED TO BE MADE

POINTS FOR DISCUSSION

The potential scope of the subject of potential "changes"—*fundamental* changes—that would aid drug development is almost limitless. But a focused approach is required if improvements are to be made. And it is essential that interested parties, ranging from patients to Wall Street, recognize each other's points of view and come to this virtual negotiating table without a fully developed and intractable "position."

To see solutions, it is important to evaluate where we have come from and where we are going. This is all dependent on the "driving forces" behind drug development. But first it is important to recognize a few salient facts. The first is that the efficiency of the drug discovery process is low. The second is that if society wants Pharma to invest in curing and treating particular diseases, society needs to find a way to make these diseases commercially attractive. There is no easy answer since drug development remains part science and part art because scientists are still relatively poor in predicting complex responses and interactions.

The Efficiency of the Drug Discovery Process Is Low

The efficiency of drug discovery is low. A snapshot of the year 2001 is worth processing and absorbing. The research and development (R&D) cost in Pharma and Biotech was about $26.4 billion. Twenty-six biologicals received new drug approval, but only nine new chemical entities (NCEs) were approved as drugs. In the chemistry laboratories of Pharma approximately 2 million compounds were synthesized by medicinal chemists at a rate of about 100 to 1,000 molecules per chemist per year. Of these 2 million compounds synthesized, about 200,000 were disclosed in 2,800 patents. This means the pharmaceutical companies thought it worthwhile to patent about 10% of what it synthesized. In addition, combinatorial chemists in companies specializing in generating new small molecules, which might have pharmacological effects, made about 3 to 10 million compounds. Companies such as Alanex, Arqule, and ChemBridge, provide chemical "libraries"[54] to the industry.

Of course, the nine new drugs (NCEs) were patented and disclosed much earlier than 2001—in the period 1995–1998. But it is quite clear that nine new chemical entities in a "typical" year is not a large number compared with the scale of the activity of producing some 5 to 12,000,000 compounds for study in preclinical tests. Why is that? It's all related to the pharmacoeconomics of drug development in the context of clinical efficacy, safety, and related regulations and legislation. What has driven and what now drives drug discovery?

Major Drivers of Drug Discovery up to 1980

Drug discovery is not only tremendously—and sometimes horrendously—regulated, but it has many *drivers*, and these drivers are changing. In 1980 the drivers were very different from today's. In order to sell a drug, a pharmaceutical company still has to convince FDA that the drug is safe and efficacious, and then the company has to convince the doctors that their

[54] Chemical libraries are "libraries" of chemical entities to be tested for any biological effect by the customer (i.e., the Pharma company). The libraries comprise molecules from all sources, including modern combinatorial- and parallel synthesis-type chemistries.

patients need it, the insurance companies that they should pay for it, and the patient that he needs to take it, and—just as importantly—to remain on it. These are not trivial things. These drivers are very important for those who want to sell drugs; therefore, the companies try to evaluate whom in the various groups they have to influence in order for their drug to reach the right people, at the right price, and for the right length of time and succeed in producing the coveted double-digit returns.

First—and obviously and understandably—there is the *medical need*. Together the doctors and patients must recognize and be aware of any medical need.[55] Next in this prioritized list would come the *profit motive of the pharmaceutical industry*. Some people do not regard this as a laudable contributor to drug discovery, but those who object to the profit motive and financial success of the industry should reflect on how many drugs have been produced by noncapitalist systems. The mighty Soviet Union had no interest in producing drugs, although one could argue that it did have the talent and means but no motive. The Soviets brought infectious diseases under excellent control for most of the population, but they only imported the best drugs and treatments from the West for a small elite.[56] A very important driver to drug discovery in the period from 1950 to 1980 was *society's need to become a healer*. For example, John F. Kennedy repeatedly made promises on the eradication of diseases, such as cancer and polio. Some cures became a reality, others are in the works, and yet others are not "doable," at least not with today's science. Finally, the mushrooming of scientific discovery contributed to drug discovery as scientists sought to have their results "translated" into practical, societal, patient benefits.

These drivers have changed in their relative importance, and they will continue to change with time. The product development and lifetime of any particular drug is approximately 27 years. The Pharma industry, state and federal legislators and legislation, insurance companies (particularly in the U. S. Healthcare Maintenance Organizations [HMOs]), and even diseases change during this time frame.

---

[55] This is not always the case. Viagra stimulated a medical need that was not recognized prior to its introduction. Pharmaceutical companies try to stimulate patients and doctors to recognize some disorders as diverse as *irritable bowel syndrome*, *social phobia*, and *attention deficit hyperactivity disorder* (ADHD).

[56] The first non dye-based drug company was in fact Hungarian. Gideon Richter is still famed for its production of steroid-based drugs. Another, Chinoin, is part of Sanofi. The Soviet Union's "Jonas Salk," Ilya Sabin, produced Sabin drops simultaneously with Salk's polio vaccine. Countries such as Finland adopted the Sabin drops in preference to Salk's injection.

Major Drivers of Drug Discovery 2000

HMOs and insurance companies are now the major drivers of not only health care, but drug discovery. The next most important is probably government and governmental agencies. Patients and their relative-backed interest groups, by lobbying government, now come ahead, in terms of their influence, of the profit motive of the pharmaceutical industry. Two recent stories put this in perspective. With all its horrors HIV/AIDS has hit hardest in Africa and, as a direct consequence, has been neglected by Pharma as much as has the development of a malaria vaccine. Obviously, these are terrible diseases for which most of the patients cannot afford the drug. The vocal and visible AIDS activist groups in the United States made AIDS a politically important disease to cure and thereby assured that any drug with efficacy will be paid for handsomely. Ironically, the development for patients in the United States who have AIDS or who are HIV positive resulted in the drugs that are just now being distributed in Africa, albeit too sparsely, but still increasingly. These are the drugs that Brazil and India decided to copy with disregard for patents, claiming an emergency, which supersedes patent laws. Although the U.S. establishment was the most aggressive in its fighting of these national policies, the United States itself would have manufactured ciprofloxacin—manufactured by the non–U.S. Bayer under the tradename *Ciproxin*—had there been an anthrax attack, using the very same logic: medical emergency supersedes patent rights.

New evidence for this comes from the introduction in 2003 of legislation against paying for prescription drugs, a move led especially by the states of Maine and Oregon. There are many examples where the focus on research has been diverted to quite rare conditions, at the expense of prevalent and pervasive indications, under the sustained pressure of "lobbying." Quite extraordinarily, this has relegated *medical need* to status as a relatively minor driver. The influence of medical doctors has also been marginalized. Patients are much more demanding of doctors whom they see as a partial, inconvenient barrier to drug access. It also seems that the altruistic ambitions of society to be a healer and a provider have evaporated. Nowadays, society emphasizes the individual's right over society's good. It may not be too strong to say that scientists and scientific discovery are almost vilified by sections of society. Society may be frustrated by the often painfully slow progress of science, but even with new technical breakthroughs, careful science is unavoidably slow.

Pharmacoeconomics plays a pivotal role. Drug development is very capital intensive and even big indications such as malaria and tuberculosis are affected. The cost means that small indications suffer, regardless of how good the science is. If drug discovery were a science-driven activity, one would expect scientists to be running drug companies. However, since Roy Vagelos[57] of Merck retired, no Big Pharma has been run by a scientist; they are all run by people who were trained in economics. The World Health Organization lists 489 diseases that need attention. What would be the motive for company executives to extend their corporate view beyond the roughly 29 recognized disease entities that are considered by their marketers to be commercially interesting with blockbuster potential?

The company's order of priorities is extremely clear. The major factors in selection of a clinical candidate in the companies' own priority order are: (1) *marketing*: the indication, the company's franchise, recent FDA approvals, labeling concerns, generic competition; (2) *internal economics*: cost, duration and effort to complete the clinical trial as compared to competing trials with the same economical potential, manufacturing capacity, cost of goods, intellectual property status; and (3) *scientific, technical, and legal issues*: the quality of the clinical candidate, likelihood of a safe, efficacious drug. The regulatory and marketing groups, and then the clinicians, can always override scientific considerations; they "call the shots."

Under current circumstances this is unavoidable. Clinical candidate selection is the most complex decision-making process in R&D. Science may have brought the clinical candidate to the table for the company's consideration and for further investment, but the strategic marketing considerations, the clinicians' views on medical need, the feasibility of the trials, and the regulatory and intellectual property issues, all come before science. If the marketing experts, who bring an entirely different form of competence to the table, demonstrate that there will be no cost-effective market for the drug, no amount of science will save the program. The different competencies of the executives in the drug company have a very different weight in decision making when it comes to the really expensive Phase III clinical trials costing $50–300 million and taking two to three years. Decisions of this caliber are so expensive and so delicate for the companies' future that they cannot be left to

[57] P. Roy Vagelos, retired chairman of the board and chief executive officer, Merck & Co., Inc. Dr. Vagelos served as Chief Executive Officer of Merck & Co., Inc., for nine years, from July 1985 to June 1994.

scientists and clinicians alone. Business strategists and economists have to weigh in heavily, even when they have no clue about the technicalities on which a drug's efficacy will depend and on which depends FDA approval of the drug.

Bringing Attitudes in Line with Needs

Should society think once more about the greater good? It is hard to discuss societal issues without appearing political. But in democracies governments and legislators look to society for guidance. There is a need for perspective and objectivity before dealing with specifics. And we have to look at cost effectiveness. Therefore, in order to preserve perspective, we do want to mention that the supply of clean water, food, and vaccines is the prime determinant of healthy human life and access to them saves many, many more lives than all these drugs we are talking about.[58]

The future of drug discovery, if it is to exploit the potential of "genomics-based, individualized medicine," assumes huge changes in society's attitudes. This is in the future, but for the present and the immediate past, it is very interesting to note that the class of drugs that has been the most lucrative in terms of "dollars spent" is the new "heartburn" drugs: the proton-pump inhibitors (PPIs). Omeprazole (Losec or Prilosec) surpassed the H_2 antagonists because doctors and patients recognized it was better—had a better therapeutic ratio—and bought the better drug despite it being more expensive and despite aggressive countermarketing from AstraZeneca's competitors.

From a business point of view, this is also remarkable. Earlier we mentioned the added value which drug companies lend to their products. Omeprazole as Prilosec costs roughly $200,000 per pound (about $440,000 per kilogram) to buy, and the raw materials cost peanuts (relatively expensive peanuts, of course). For comparison, an F-18 Hornet fighter only costs around $400 per pound (about $880 per kilogram)!

---

[58] Modern science can present new dilemmas. If scientists did discover a vaccine for HIV, it would present a very difficult situation; it could not be delivered *en masse* as in earlier vaccination programs. Nowadays, vaccine programs have to be voluntary. Vaccines such as that for hepatitis-A are expensive and elective. They are not always covered by insurance programs since they are preventative.

The cumulative sales of Losec are currently reasonably estimated at about $40 billion. But the savings to society are about $85 billion because the number of gastric ulcer surgeries required has decreased dramatically by 75%. This is extraordinarily important if you consider that these surgeries, as we previously mentioned, accounted for 80% of operating theater capacity in the United States. In addition the complications with added costs of surgery are avoided.

All this investment in heartburn care should be noted in the context of heartburn being in many patients a very self-inflicted condition. Better diet with less stress and late-night pizza may have the same effect as PPIs and gastric surgery.

Bringing the FDA in Line with Needs

How do the people in the FDA, and the international equivalents, look at their role, and how do they judge specific drugs submitted to them for approval? A significant problem the FDA has is that there are many "me-toos" submitted as companies try to develop their own drug in a particular class for a particular indication. The companies see it as a way of generating profits, through establishing a new market share, and it is also seen as a safe way to introduce a new drug to the market. The company with the "first-in-class" pioneer drug has already clinically validated the target. Moreover, it has developed the market and often made the mistakes that enable the second and third drugs entering a market to make higher profits. But these new drugs in a class may not represent a real improvement, and they do not fulfill a real medical need. It is, of course, nice for physicians and patients to be able to try several SSRIs before opting for a completely different class of antidepressants. But does one need seven of them? Of course, since they could be patented and they are chemically slightly different, some patients may tolerate the side effects of the seventh SSRI better than of the other six, but all SSRIs expressly affect the same drug target with the same mechanism of action. The 30% of patients with depression who are treatment resistant have no better benefit from the second to the seventh SSRIs as they had none from the first. But no one is thinking about the patients, just market share. This 30% would need a completely new drug, and making that is a much more unsafe proposition for a drug company than making the fourth, fifth, sixth, or seventh patentable SSRI.

The FDA is looking at the drug's treatment potential and is seeking drugs that would really advance medicine in some or another way. This could be either by the drug being very much more efficacious, or by eliminating a very significant side effect of existing therapy and medicines, or both. For these drugs, as was the case for the drugs submitted as antivirals against HIV, they give a "priority review" where, in extremely urgent cases, some standards of proof of efficacy and even safety may be relaxed. However, for the majority of drugs submitted for review, they give a standard review since the drugs have similar therapeutic qualities to those of an already marketed drug.

What could be done to improve the FDA's mission? The FDA is responsible for regulating medicinal therapy, the approval, labeling, and marketing practices for drugs, but not the practice of medicine—that is, how physicians prescribe the approved drugs for what indication and in which dose and under what type of surveillance or monitoring. The FDA also regulates diagnostic testing and the approval of medical devices. Its stated purpose—as a nonprofit governmental agency—is to protect the public as patients, not to invent new drugs, or even point to such possibilities, but simply judge if the proposed treatments submitted to them are, according to our present scientific understanding, efficacious and have acceptable risks.

One aspect of health care that does not fall completely under the FDA's control is surgery. Compared with medical care, surgery is very much under the control of surgeons as represented by their various academies and organizations. They determine accepted surgical procedures and negotiate with insurance companies and Medicare/Medicaid for payment.[59] This can have a severe effect on drug discovery if only because the cost of surgical procedures comes out of the same budget as medical care and prescription and other drugs.

Why might surgery be more closely monitored by regulatory authorities? There are two illustrative examples. A new procedure called lung volume reduction surgery (LVRS) reportedly helps emphysema patients who have severe difficulty in breathing and exercising. The procedure, together with hospitalization bills, costs some $70,000 per patient more than traditional medical treatment, which itself is some $13,000 per year. Should insurance companies and Medicare pay for this? "Of course," might

[59] The FDA does have a role in analyzing data from trials involving surgical procedures. Of course, all of us understand that surgeons time after time find unexpected problems while performing what surgeon and patient thought to be a routine surgery, and we all rely on their ability and experience "to cut themselves and us out of trouble." Fully determining by regulatory means what they should do would be counterproductive and potentially dangerous. We do not want automaton surgeons. But the drugs and recommended doses to be used by the anesthesiologists, and so on are clearly laid out by the FDA and are being modified by medical need during the procedure.

be considered a rational response. In fact, the surgeon who invented the procedure, and who clearly detects a benefit from the procedure, has been reported to admit to encouraging his patients to sue whenever there was any hesitancy to pay.[60] The patients won the lawsuits. But should the surgery be performed? This is more controversial. There is little evidence that the procedure is beneficial. The most rigorous analysis of the data showed that only a subgroup of patients—those who had upper lung emphysema and had difficulty in exercising—benefited. They lived longer and could exercise more. But against this is a 17% mortality rate from the procedure and the trauma of the surgery itself. Should more trials be conducted to determine who might benefit? "Of course," might be considered a very sage point of view. But, again, surgeons are allegedly not keen to participate in any new trial since in order to determine if the procedure is in fact beneficial, some of their patients would be put in the "control group" and would *not* undergo the surgical procedure. Surgeons are loath to deny their patients procedures that *may* benefit them.

This was very acutely seen in an important long-term trial comparing a surgical procedure for certain strokes called carotid artery bypass surgery, also known as *endarterectomy*, with a medical procedure known as *taking aspirin*. Aspirin was known to have potential protective attributes by virtue of its anticlotting action. The medical question was: "Does the surgical procedure protect patients from a second stroke more than preventative aspirin?" Dr. Henry Barnett of London, Ontario, Canada, conducted the multicenter trial. If a patient presented with a stroke and was a candidate for endarterectomy, the hospital would call a central number and the patient would be randomly assigned (1) endarterectomy or (2) aspirin. So many institutes withdrew from the program, with surgeons at the institutes electing to operate on every patient, that the trial had to be continuously expanded to more and more institutes, and even new countries (Japan, etc.), to gather enough data for statistical significance. The result? Cheap aspirin is just as good as expensive surgery. Detractors of endarterectomy claimed that surgeons were too keen to perform surgery, which was not very dangerous and from which patients recovered and showed improvement, and for which surgeons could charge large fees. Detractors also noted that patients recover from such strokes without any surgical intervention. Surgeons, who did not want the procedure to be discredited or become nonreimbursable, organized their own lobby groups of celebrities who swore by the efficacy of the procedure.

[60] As reported in the *New York Times*, August 17, 2003.

If the FDA allowed pharmaceutical companies to campaign so vigorously without data to back up their claims, there would be a legitimate outcry. But society should know that all the money that might be projected to be spent on expensive surgeries far exceeds the money that might be spent on prescription drugs aimed at the same, in this case, aging population.

Bringing Legislation in Line with Needs

Legislators should be aware that lobbying, whether by patient advocacy groups or Big Pharma, doesn't always serve the public good. They must also come to terms with the fact that choices have to be made about which medical and surgical procedures are supported by government agencies, specifically Medicare and Medicaid. Who should be allowed to choose if someone should have a procedure that occurs, say, as part of the normal aging process, or as a result of self-inflicted injury (be it smoking or poor dietary control)?

The most important legislation that could transform and improve health care would be a revision of Medicare and Medicaid programs and legislation related to insurance coverage. For the moment this is outside the scope of this book, though we are tempted to suggest that insurance companies should be encouraged to look more favorably on prevention. Indeed, cholesterol-lowering drugs, prescribed to prevent—or at least reduce the risk of—cardiovascular disease, had sales of $8 billion in 2002 and the costs are reimbursed by insurance programs. This is because long-term trials by Merck and Warner-Lambert-Pfizer showed clear benefits in terms of prevention, and thus savings, within a few years. And if individualized medicine is to "take off," then the insurance companies and the FDA must change, and patients must be protected and securely insured. Meanwhile, perhaps it is up to the politicians to keep clean water and vaccines in mind?

part

II

From Basics to Bedside

chapter

9

TARGET-BASED DRUG DISCOVERY: PART II

THE EMERGING DIVERGENCE OF ACADEMIA AND INDUSTRY

Scientists can work with industry either from the secure confines of academia or by joining industry. Since a large number of research students are at least tempted toward a career in Pharma or Biotech, it is important for scientists—the students and their mentors—to understand the differences. From such understanding come insights into successful collaboration on drug discovery.

One of the major reasons research scientists are drawn to industry is that they see better paid former peers and colleagues working in well-equipped, well-funded facilities and a possible escape from an often cramped academic environment with pressure to compete for laboratory space, funding, and continued employment. But nothing is that rosily simple.

There are major differences between academia and industry, but these are seldom recognized in contacts between academia and industry because people who want to collaborate tend to emphasize the similarities rather than focus on what divides them. Needless to say, the academic and the industry scientist have the same formal academic training; they were often classmates. But from there on, three years after their paths diverged, there are huge differences, and they are often not emphasized. But if a scientist who has left the cloisters of academia for the greener pastures of industry forgets the differences, he or she is in for surprises. Where do academia and industry diverge?

Academia is "science driven"; it's good enough to say when advocating or promoting someone: "This is good science." The research project focuses on a problem, and a senior scientist—or "principal investigator"—manages a project. The motive is the science[61] of discovery, of distinguishing between different possible phenomenological explanations or biological mechanisms. In an academic environment, it is possible and desirable to be a very specialized expert on a given gene or gene product, be it an enzyme, a receptor, or some other protein. The value of this is not to be underestimated; the molecule may be a key to a particular disease. But if it isn't, the molecule itself and the scientist[62] are of only passing interest to the Pharma industry. However in academia this may be a "celebrated" protein because of its particular complexity of structure, rapidity of its decay, and so on; the more extreme and esoteric, the better it is for a scientific career and for science itself.

There is a big difference between the choices of topics for academic scientists and the choices of industry. Industry, whatever it might say, is profit driven and market focused. Ph.D.-level scientists in a pharmaceutical company work in line-management mode. That is to say, a scientist may be hired to do "all" receptor work, not, say, just histamine receptor work. In fact, if you are the histamine receptor guy and the company moves out of histamine receptors (because there are too many and the company has lost interest in allergy and in acid stomach), your immediate career may be in big trouble.

The company, especially Big Pharma, is working on many projects, and, typically, a scientist would be assigned to a therapeutic area rather than to a gene or gene product. What is potentially very stressful to the scientist at the personal level is that in the drug company, every project, every single day, is compared to every other project, not only those within the company, but also to all the projects of all other companies. Whatever you are working on might be "licensed in," exchanged for other projects, or replaced by the competition's rival project as the competitor is acquired or merged with. Your pet project may be handed to someone else as your project is merged when your company is "merged." Companies are always

---

[61] Science: Middle English from Old French (= French) ***science***, from Latin ***scientia***, "knowledge, science," from Latin verb ***scire***, "to know," which probably meant originally "to separate one thing from another, to distinguish," and is related to ***sciendere***, "to cut, split, cleave," and Old English ***sceandan***, "to divide, separate" (from Klein's *Etymological Dictionary of the English Language*).

[62] Scientist: A hybrid coined by English philosopher William Whewell (1794–1866) from the Latin ***scientia*** and the Greek suffix **–ist**.

likely to establish internally new priorities between projects, and your project may suddenly have support withdrawn. This is not the case in academia. Hospitals may merge, but a Harvard-Berkeley merger is way off. Even grant review committees only need to compare a handful of projects within the same area.

The stockholders of a Pharma company are totally indifferent to whether the company makes its dividends in dermatology or in cardiovascular disease. But they do care if the company is beaten by another company they did not invest in. Or worse, they care if computer industry profits are better than Pharma profits because Pharma companies are not run efficiently (i.e., Pharma is obliged to close marginal or unpromising projects fast to cut costs, and to focus on projects that more assuredly will lead to drugs).

One in academia never talks about the "cost of opportunity," although one could, while it is disclosed all the time in the industry. If you stop this project with slow progress, the new one may go faster and better, but as an expert on worms you are not likely to consider if your research would be better should you switch to ostrich.[63]

A scientist is remembered and acknowledged for his or her contribution to the science, no matter when or where such discoveries were made. The Nobel Prize usually comes, if ever, many years after the research was performed, published, and the results widely understood and used by others. In contrast, very few company executives actually are celebrated for the drugs that were made and launched by the company during their tenure, or, indeed, blamed for the drugs that failed, which also came out during their tenure. The product development time of 8 to 14 years seems infinitely long compared to the tenure of most executives. So you are never blamed or praised only for what you have done. In fact, it is a huge problem in this industry because compensation schemes cannot be directly related to the actual achievements. Achievements are measured in stock price, on sales and profits, and mergers and acquisitions. Hence, no Big Pharma company is run by a scientist anymore. In any other industry it doesn't take 14 years to develop the product. And the average executive remains only 4 years at his or her position. To reconcile the short tenures and the long development times for a drug, executives in R&D in a Pharma company are judged by the number of clinical candidates generated, not by the number of drugs on the market.

[63] Although it should be noted that the 2003 Nobel Prize winner Sydney Brenner was famous for advocating the original work in a worm (*Acaris* rather than *C. elegans*) that led to massive advances in knowledge about development and cell fate. He has also implored the zebra fish community to switch to puffer fish because the genome of puffer fish seems to be wonderfully devoid of lots of inactive regions. His argument is that progress would be faster, but zebra fish are easier to farm than puffer fish might be.

Analysts study companies' pipelines and calculate the probabilities of successful clinical trials in preparing advice on their own investments. Investors tend to have a much shorter-term view than industry managers and scientists can afford to have. The influence of investors on drug development decisions should not be underestimated, and the influence is not always positive.

Academia—Pharma Industry Discussions Are Sometimes Difficult

Scientists in drug companies and academia may have the same experience and education in their first year of employment, but their vocabulary, point of view, work culture, and values diverge sharply thereafter. About 1.6 million people are working in the pharmaceutical industry world-wide, of which only about 10% have a Ph.D., so there is much more to this high-tech industry than just science. Clinical development, production, distribution, and marketing have much larger departments than the science-driven preclinical department. However, it should be noted that the Pharma industry employs almost 100 times as many chemists and biologists with Ph.D.s than the University of California, Berkeley, so the intellectual firepower of Pharma, together with its capital strength, makes it a very powerful factor in world research. In the area of life sciences, the Pharma industry outspends government and private universities by 4 to 1 or more. When the U.S. government through the National Institutes of Health (NIH) decides to try its prowess in drug development—as it is now trying to do—it is planning to spend much less than any of the top 10 large Pharma companies alone.

Industry often goes to academia for consultation and to bring in expertise. The expert scientists are consulted on specifics, not for any decision making related to the company's investment in a particular project. Whether through deference to the scientist's prestige, or just a symptom or side effect of secrecy in the industry, the experts are excluded from decisions, especially the decision to close a project that the scientist might be advocating or opposing on scientific grounds alone. The expert's advice is not needed or heeded. This is certainly so for preclinical experts, that is, chemists and biologists. Clinical "authorities" imported for advice are listened to somewhat more. They may explain what they think their physician colleagues will do with the drug and suggest what measures and side effects will be most hotly discussed by the FDA and practicing physicians. Since they are the "opinion makers" of this field, they are listened to most by the marketeers. If they have some scary thing to say about a potential product, they may contribute to abandoning a project for fear of liability, but even these greatly influential clinicians cannot start a project, or decide how to develop a product, and when and where to market a drug. The Pharma company is balancing factors such as changing market conditions, a competitor's new product, the FDA's new criterion for evaluating and measuring improvement in, say, multiple sclerosis, or, worse, the FDA's new indecision about how to measure improvement. The last scenario would put every company that wants to make a multiple sclerosis drug in limbo. Good reasons for closing a project may make sense for the company, but they might not make sense for the scientist who was relying on the project continuing. A large number of projects are closed before reaching clinical trials or in early clinical trials. Budgets are based on the knowledge that preclinical research is a lot cheaper than clinical research and also that attrition rates for the industry are known to be fairly comparable between the large Pharma. In other words, the closing of projects and programs is expected, budgeted, and scheduled. And no one bothers to explain all this "programmed project pruning" to the scientists inside or outside the company.

Target-Based Drug Discovery or Finding Clinical Candidates in an Academic Setting and What to Do When You've Found One

Scientists do not have to go into industry to contribute to drug discovery; it can flourish in an academic setting. Indeed, scientists may find academia to be a safer environment in which to explore a career in drug discovery. Joining the industry is most likely to be an irreversible decision; hardly any scientists come back to academia (see Figure 9.1). However, if scientists want to contribute effectively to drug discovery, they should do it in an institution that has strong traditions in chemistry and not only in biology. Medical schools that have emphasized biology at the expense of chemistry really don't stand a chance in this era. They will be asked to do clinical trials— a qualified "work for hire"—but they will not, as a rule, discover drugs.

Nowadays, partially because of pressures from society looking for cures, much of academic research is linked to investigation of disease mechanisms. The National Institutes of Health (NIH), and their component institutes such as the National Cancer Institute (NCI) and the National Institute of Mental

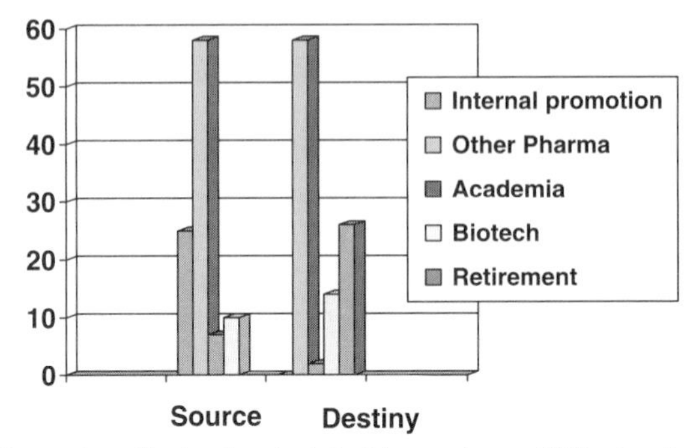

These data show from where the top hundred decision makers—CEOs, heads of R&D, and heads of marketing for the top 10 companies in 2000—are recruited and how they leave: (1) promotion from within; (2) hiring from competition (very common); (3) hiring a few from academia, which companies might regret later; (4) a few from Biotech either intentionally or, often temporarily, as a result of buying the company. Regarding attrition: Where do the top hundred go? Most go to other Pharma (to get a raise or more power), some go to form their own Biotechs, very few go back to academia, and the rest retire.

Figure 9.1 Movement of Decision Makers in Pharma Research: Very Few Scientists Return to Academia

Health (NIMH), are the main source of academic funding through their extramural funding budget of several billion dollars (about 20 to 30% of what Pharma spends on the same topics). In addition, over recent years the Pharma industry has sponsored research laboratories in academic institutions, especially those with medicinal chemistry know-how. Of the nearly 2,000-chartered organizations of higher education in the United States only a handful of universities have a medicinal chemistry program.[64] Academic research devoted to drug discovery is entirely "preclinical" since there is no conceivable way that an academic institution can afford financing or accepting the liability of clinical trials.[65] On top of this, the undeniable fact is that clinicians at medical schools would want to be paid by Pharma at industry rates rather than at university rates if a project proceeded to the clinical trial stage.

[64] For example, the Scripps Research Institute (TSRI), University of Kansas and the Mayo Clinic and, internationally, University College London.

[65] Universities may, however, carry out very small pioneering studies on an investigator-sponsored investigational new drug (IND) procedure with FDA approval.

Modern drug discovery is "target-based." No program will be any better than the protein target with which it starts. Scientists need to identify, describe, and characterize in intricate detail a molecular target for which they will make the best possible ligand[66] as specific, as selective, and with as high affinity as possible. This ligand becomes the clinical candidate: the "fruit of the preclinical research." It is the compound that is intended for "entry into man" (EIM). All of this can be first studied in an experimental system *in vitro* (i.e., in a test tube and not in a whole animal); only later does one worry whether this works *in vivo* (in a whole animal, in man). This represents a big shift from how drugs have been discovered earlier, when candidate drugs were directly coming from studies in animals (i.e., in animal models).

If you are in basic pharmacological research and find a potential key target, or, even better, a clinical candidate or chemical to treat a disease, you may be tempted to: (1) take and sell your idea to Big Pharma; (2) seek to work with your discovery in Big Pharma; (3) form your own Biotech company; or (4) publish and allow free use of your findings by scientists in academia and Pharma alike. If you choose either the first or second options, always work with or go to a company that has already developed another drug for the same indication, no matter what offers you might receive from other companies eager to enter this area. The reason for this is simple: the company with the most experience in a field has the best clinical development experience in this disease area, the marketing strength and expertise, and the facilities and resources most suited to work with your discovery. They may only give you 5% royalty instead of the lucrative 10% royalty offer from those who only with your help could enter this therapeutic area, but "5% of something" is always better than "10% of nothing." Even big companies are beginners in the clinic when it comes to entering a new therapeutic area. If you choose the third or fourth options, then there is much more to consider. Many biotech companies have been formed on the basis of discovering one target and on finding one clinical candidate. Investment is attracted to the idea of the profit potential of a significant new drug aimed at a new target, and thus it is supported with

---

[66] From the Latin ***ligandum***, the gerund of the verb ***ligare***, to bind. A ligand is any molecule that forms a complex, interacts, or binds with a target (usually a larger protein molecule of a cell), but is most usually reserved for describing a molecule which will interact with a receptor either to enhance (in which case the ligand is an agonist) or depress (in which case the ligand is an antagonist) the target receptor's signaling. For example, the renowned beta-blocker is a receptor *antagonist* used to treat hypertension. It is a "ligand" to the beta-adrenergic receptor—"the target." This *ligand* blocks the target, preventing the natural ligand, *norepinephrine* or *noradrenaline* binding, and thus the antagonist prevents "signaling via the target receptor" that would normally increase heart rate and contractile force. The beta-blocker thereby acts to reduce blood pressure.

large injections of funds into the Biotech companies.[67] The risk of option (3) is great, but the potential rewards were too tempting over recent years as hundreds of Biotechs sprang up sometimes without really robust business models. The venture capitalists have learned a lot of sobering facts about attrition of Pharma projects, and there is a considerable push to demand some kind of proof of concept experiment before financing new ideas. This raises the bar considerably for the academic researcher. He or she now has to team up with people who can help with some *in vivo* proof of principle before even the first small financing. This is a huge change from getting funds based on molecular biological findings that was prevalent in the late 1980s.

Option 4—publishing and allowing free use of your findings—may be the least helpful of the choices. If a scientist does choose this route, he or she should definitely patent the discovery, for, it bears repeating, if a discovery is not patented, no one will have any incentive to invest in the discovery and it is unlikely to foster drug discovery (see Figure 9.2). R&D teams in Pharma companies are much more likely now to publish their results in scientific journals but after first having protected their company's investment with relevant patents. Earlier, pharmaceutical companies kept quiet about their research, following old-fashioned competitive doctrines. Now it is more widely recognized that if your company has in fact made significant discoveries on the path to developing drugs in a therapeutic area, it is good,

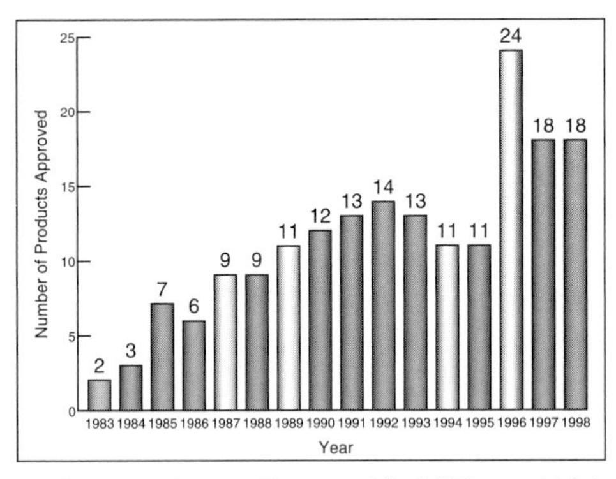

If you forgot to patent or, for example, you discovered that lithium, which is hard to patent, is the best drug, then there is orphan drug status to help companies help patients.

Figure 9.2 FDA Approved Orphan Products, 1983–1998

[67] History may already have questioned the wisdom of this, and a better approach might have been for scientists to form the equivalent of a consortium of Biotechs with shared resources, a sort of Universal Studios for the industry.

effective advertising to publish this fact. Your competitors are more likely to pursue an alternative noncompetitive route if they see that you are ahead in your research. This policy can also facilitate collaborative research between a drug company and an academic institution where the goals are complementary. It is often emphasized in debates about whether or not to publish that the company will be able to attract and retain more talented scientists if it publishes. Most importantly, investors have become more sophisticated, and, as mentioned earlier, the analysts are comparing "pipelines of drugs in research phase" for judging the potential worth of Pharma companies. If you do not want to have your share price hammered, you have to disclose how your research is going, at what you are aiming, when you will enter into clinical trials, how long they will take, what the FDA's position is on this, and so on. While the actual chemical structure of the drug is kept secret until it becomes an approved drug, all other information is packaged nicely for analysts to impress them. You cannot persuade them to believe you and yet be fully secretive. It is a new big problem. Analysts have PhDs and MDs; they are no longer bankers only, overwhelmingly because the money in Pharma is measured in trillions of dollars. When a Phase II trial for a new class of antidepressant—a so-called NK-1 antagonist—failed at Merck, the company's valuation fell roughly $5 billion. They had not lost any income; the candidate drug was still four years away from generating any revenue. Merck followed up with a new Phase II trial six months later, yet its stock valuation did not go up. Another year later it closed this project. Meanwhile, fluoxetine, the most popular antidepressant, became generic after Lilly's Prozac lost its patent protection, making the market for new antidepressants harder to enter. The stock market valuation is very sensitive to failure. Everyone believes failure. Success has a slower effect on pharma stock prices. Speculation is more effective than success.

Adjusting Academic Truths to Industry Values or Putting the "I" in Industry

Academic scientists are subjectively critical of the pharmaceutical industry because of the financial structure where profit must come before scientific understanding. But it is not the authors' intention to add ammunition to that point of view, even though we are providing often critical insights into the pharmaceutical industry. Furthermore, pointing out that "everything is not perfect" does not represent an attack on the industry. Objectively, scientists should be extremely impressed with the Pharma industry because it generates drugs that alleviate suffering. It bears repeating that noncapitalist systems have been very poor at producing these drugs. While the elite behind the Iron Curtain could have access to sophisticated medicines of the West, and while it is also very true that the Soviet bloc's vaccination-based health programs were in many ways equal or superior to those in the West, general medicine, facilities, and resources were poor. For example, if a child had epilepsy, he was given the one available antiepileptic. It was a highly sedating barbiturate, and, while it largely controlled the seizures, it meanwhile impaired learning. Nobody behind the Iron Curtain could afford to care about this side effect. Without the inherent competition that capitalism with all its faults imposes, and upon which this Pharma industry depends to be a $300 billion a year business, new drugs won't come forth. If scientists are invited to collaborate with the pharmaceutical industry or to join it, then this should be regarded as a positive step, because the industry benefits from having highly trained and creative scientists in its midst. In effect, scientists can simultaneously be both on the side of academia and the drug companies, and this is being more and more appreciated. Most companies award prizes to their scientists and are proud to let them give lectures together with the best of academia. However, this notwithstanding, if as an industry scientist, you "made a drug," that is, if you were involved in a project that really produced a medicine, then you are in an infinitely stronger position as a Pharma researcher than if you had been invited to deliver a plenary lecture at a prestigious symposium.

If you are a senior academic scientist and are approached to join a pharmaceutical company, you might indeed be tempted and take the plunge. This often-irreversible move may indeed be worth it despite its provoking jealousy or hostility in your former colleagues who frown upon a career in industry as being "academically impure." These same former colleagues will be first in line to ask you for research resources, consultancies, congress support, and the like within weeks of your industrial arrival. Your former colleagues should be more sympathetic, for after the courtship, reality checks in. Your arrival will be greeted with a note from your new bosses as follows:

Dear Professor McSmith:

Welcome to Amerga, Inc., PLC, BV, AB, SA, Oy (etc.). We are delighted that you have decided to join our company in Nowhere in the heart of Europe/the Garden State of New Jersey.

Your first and most important task (among your 1,001 administrative ones that we forgot to outline earlier) is the SELECTION of winning drug targets.

No company can be successful without them and your compensation scheme makes you a valuable participant in our success.

You are not to waste valuable screening efforts and medicinal chemistry resources on lousy targets.
You are not to take risks with the stockholders' money or your colleagues' compensation; you are to come up with breakthrough medicines.

Thus, you must have a balanced portfolio of targets.

Until your next evaluation, we remain

Yours most and verily sincerely,

V. Impersonal, MBA
Chairman & CEO

Targets are the key to drug discovery. What are validated drug targets? What is their value?

After all the preclinical efforts, clinical candidates and targets have to be validated in humans. The company goes to the FDA and asks for an **IND**—Investigational New Drug—license that permits the drug to "enter into man." Each clinical candidate has to be tested *in vivo* to obtain a "Proof of Principle." More than 30% of all clinical trials supported by Big Pharma in the past five years[68] were Proof of Principle (PoP) trials. Many times the industry is spending $10–30 million to obtain a PoP in humans, often using "imperfect" but safe compounds that are clearly not going to be drugs. This might be just because of inconvenience of delivery, and so on, but if they are safe enough and selective enough for their target, then they may give very useful information: is the intended action beneficial? Is it strong enough to matter? Are there inherent side effects with using this molecular target, for example, is there a target-based toxicity that will remain even if we make a thousand new and better molecules to this target? The Proof of Principle is thus a key step for continuation of the efforts. Thus, the first clinical candidate might have had to be delivered too often (e.g., four times daily) and inconveniently (e.g., unpalatably huge tablets or by subcutaneous injection) or it might have had too short a shelf life even to make it into pharmacies' inventories. But if the trial shows that the candidate "hits" the right target, and that hitting this target would have the desired therapeutic effect, then the principle has been proven, or at least demonstrated. Of course, a negative or inconclusive trial could invalidate the target or, at best, simply banish the clinical candidate from the reckoning.

---

[68] 1997–2002.

A successful Proof of Principle, and the validation of a target as useful for treatment, is always dependent on medicinal chemistry. (There might indeed be new cases when a therapeutic antibody can provide it—or even the antibody may become the drug—but this is still rare and is not the mainstream of Pharma research.) The medicinal chemistry, by which the compound is found (i.e., designed and synthesized) to be put into these trials, is often a prerequisite for target validation. In other words, the ligands the medicinal chemists make are the best way of validating a target for future pharmaceutical use. Modern genomics and proteomics increase the need for medicinal chemistry as they provide new targets for potential ligands.

However the Proofs of Principle trials work out, as the new director of the R&D program, you will, of course, be judged on results, not effort or ingenuity:

Dear Professor McSmith:

Thank you for your contribution to our company's success this year (we, of course, know that the success of this year's marketing launch was the work of your predecessor who left 4 years ago to our competitor but we cannot send him a bonus, as you may well understand).

That being said, we appreciate your work and willingly reward you with a bonus. The bonus you were to have received has been split in half because you have failed to in-license a sufficient number of good targets from Biotech (*expensively and with royalties, but nevertheless . . .*). We take this opportunity to remind you that we really do not care where the targets come from as long as they are good and drugable in large indications.

We wish you Happy Holidays!

Sincerely,

V. W. Impersonal Jr., MBA
Vice Chairman & COO

PS: We are looking forward to judging your selection of clinical candidates, which are the fruit of all preclinical research. Your next bonus will reflect your understanding of this principle.

The company will judge you on the selection of the clinical candidates, because they are the products of all the preclinical research for which you are directly responsible.

chapter

10

"DRUGABLE" TARGETS

GENOMICS AND DRUG DISCOVERY FOR PATIENTS

What exactly is the promise of genomics—the study of all the genes of an organism and their function—and what is proteomics—the study of the gene products (proteins) and their function—in the development of better medicines?

GENOMICS AND "DRUGABLE" TARGETS

Scientists are working on the basis of the premise that they will find high-quality, validated targets from new knowledge arising from the human genome project. Some diseases are obviously linked to genetic variations, mutations, or so-called polymorphisms,[69] and may be detected through family-linkage studies, where the occurrence of disease is traced through

[69] Literally "multiple forms" of a gene that exist in the population, and some of which forms are associated with a disease or vulnerability to a disease.

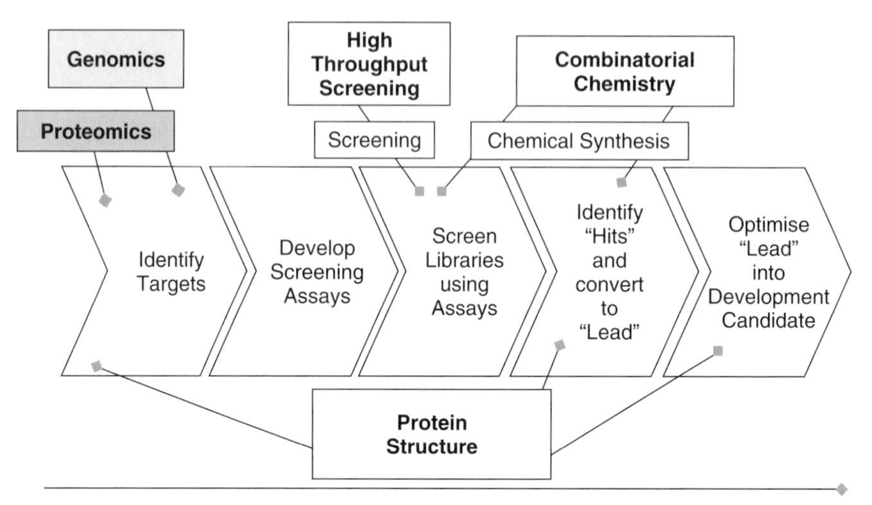

The figure shows the steps along the path of the process of drug discovery. Genomics helps with target identification but not with screening assays, chemical libraries, finding hits, and converting them to leads. In the future, genomics may help with the "stratification of the patients." In other words, it will answer the question of whether the patient has the target or has abnormal levels of it, and, in addition, in some instances, whether the patient has another condition—another target—that would make the drug inappropriate, i.e., a "molecular contraindication." Toxicogenomics can affect lead optimization.

Figure 10.1 Genomics and Proteomics Aid the Discovery Process

families. Other genetic diseases are not so obviously inherited and manifest themselves because of, for example, fragilities in the genes that are not always detected or expressed as disease. Such fragilities may become more prominent between generations.[70] They might also require specific environmental triggers, for example, exposure to a viral or bacterial infection at a critical time, an allergy, and so on, to cause disease. In the absence of such

[70] Such as is seen in so-called triplet repeat disorders (TRDs) where for unknown specific reasons, a triplet of nucleotide base pairs—such as CGG in Fragile-X syndrome—expands between generations. There might be 5 to 50 repeats in the general population, but when this expands to 45 to 200 repeats this can become a genetic catalyst—a "premutation"—and if over 230, and up to several thousand repeats are in the male offspring, the boy will have mental retardation that becomes apparent in late infancy.

triggers, the disease may never manifest itself and the genetic defect or vulnerability remains "silent," but it will be inherited in some fashion.[71]

In some complex diseases, mutations in one or several of the genes lead to slight variations in the form, shape, and chemistry of the gene products, making these proteins usually less efficient at their biological tasks such as their interactions with natural ligands. These mutations cause "loss of function." Others cause "gain of function." Neither of these types of mutations is healthy as cells have not been prepared for these aberrant functions or disruptive levels of function, and which may now cause disease manifestation. Diseases, such as cystic fibrosis (CF), arise because of a small variation in a single gene, which, in the case of CF, leads to a catastrophic variation in a protein that indirectly controls the water transport in cells, resulting in severely and, eventually, fatally congested lungs. Most difficult will be the understanding of diseases with multiple genetic components where two to three genetic variations must be present for the disease to become obvious. Most scientific advances have focused on genetic diseases caused by one faulty gene such as Huntington disease[72] or cystic fibrosis (CF). Indeed, scientists may be guilty of looking for single genetic changes (mutations) for complex diseases. The list of diseases with complex genetics would include many of the mental health diseases such as schizophrenia, where simple rules of heredity or receptor function are inadequate to characterize the disease at the molecular level.

With the full sequence of a human genome, which was completed in 2002, as a resource, scientists will be able to map variations of genes to disease states, and the gene product of the affected gene becomes the target of choice. In addition, having found and identified a potential target, proteins in the same class as the putative target may become targets themselves. Known ligands are potential starting points for a "proof of principle" evaluation.

In theory, all Big Pharma companies have the ability to exploit equivalent leads. In practice, however, the difference in success rate between

---

[71] As will be seen in later chapters, this is an important but insignificantly discussed aspect of genetic testing as part of health insurance schemes. Because some of these types of genetically identifiable mutations do not carry risk before exposure to known or unknown environmental factors, it should be with great caution that a physician, an insurer, an employer, or society should declare someone as "less insurable" because of a genetic abnormality. The mutation may be easily found by genotyping, but without any detectable manifestation, society should protect the integrity and confidentiality of individuals' genotype data. Information that should help the patient and the physician should not penalize the patient through its discovery.

[72] Formerly called Huntington's Chorea since this was how the disease was described by the Long Island neurologist George Huntington in 1872 (ref: Koehler, Bruyn, & Pearce: *Neurological Eponyms*). The current trend is to drop the possessive apostrophe and, in this case, the politically incorrect "chorea," which comes from the Latin for "dance," since sufferers exhibit jerky uncontrollable movements. It is also a triplet repeat disease (TRD) with a sequence of more than 36 CAG repeats required for manifestation.

companies manifests itself in many small differences between them all along the drug discovery process starting with their different chemical libraries. Will there be any kind of compound from the library that will serve as a "hit" from which one will be able to start a medicinal chemistry effort? How likely is it that one can find a compound or several compounds of moderate affinity to serve as starting points?

The major worry for a company in high-throughput screening (HTS)-based drug discovery is that there will be no hits to a new target from any of the compounds in its chemical libraries. Once a company has a compound that binds even ever so slightly but specifically to a target (i.e., even with low affinity), this is most likely a "drugable" target if they are given enough time and resources, such as 12 to 18 months for a team of 6 to 12 chemists. From even a faint lead medicinal chemists will most often be able to improve the tightness of binding and the selectivity of the drug candidate. If a company has no "hit" of its own its scientists may use other companies' published hits—even approved drugs—as the starting point from which to improve characteristics and to develop its own compounds.

One must be aware, however, that good chemists at the rival company with the established drug to this target have probably patented every conceivable compound around the drug molecule as well as the chemical intermediates that led to this molecule. If the chemists at the company looking for the lead come up with one that the rival has already patented, even if the patent was not yet in the public domain (18 months), all work might have been in vain because infringing on a chemical patent is clear cut, and not permissible.

Owing to two main factors this situation arises reasonably often. First, the biological target has a very tight "spot" or "cavity" and only one kind of molecule will fit in. The literal space and room for patenting is restricted. Therefore it is not that rare that a chemist will arrive at similar molecules to those made by chemists at the rival company. Second, a lot of medicinal chemists come from a relatively few outstanding schools and may have certain ways of solving a problem. If their pathways of discovery overlap, the result is patent infringement. If you wait 18 months for the patent to issue and be public, you will never know how many others are chasing this patented molecule and when they started. Clearly, the best way forward is to have your own "hit" or even better "hits," that is, to have a really diverse and large library of millions of diverse and unique (i.e., not so trivial) compounds and the ability rapidly to screen these until hits against the target are found. If there were a good objective measure of the diversity of companies' libraries, that is, how likely a given company's library would produce a hit for a new target, it would greatly influence drug companies' valuation on Wall Street, as much as any key technology possession

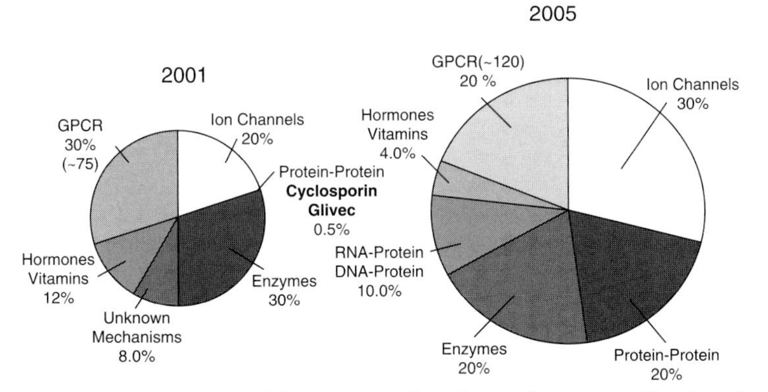

The projection is that the number of drug targets will continue to increase and the diversity of targets will also improve. From a base of about 250 targets in 2001, the number of targets has increased to about 600 in 2005. By working on these targets, one would reasonably expect around 1,200 NCEs giving rise to about 15,000 new drugs by 2010.

Figure 10.2 The Number of Drug Targets by Category, 2001–2005

influences valuations of other industry companies. The promise of genomics-based drug discovery is, therefore, that new clinical candidates will be found for the new classes of targets identified by genomics, provided that chemists will be able to match the expansion of genomics produced targets (see Figures 10.2 and 10.3) with relevant leads.

The prognosis for all this great effort across the whole industry is "cautiously optimistic." By 2010 a reasonable projection is that a biological Proof of Principle will exist for about 600 or more targets. Of these, about 100 to 150 will be validated and functionally be able to be targeted, with an expected 1,200 new chemical entities (NCEs), and about 15,000 "drugs." There are two major ways by which the number of drug targets is expanding. First, the examination of the human genome with the resultant definition of new genes and the establishment and validation that the proteins they encode are involved in disease processes. Second, the identification of molecular pathways where the whole cascade of molecular events in disease processes are elucidated together with all the participating protein components. If the association is made between a protein member of this "molecular cascade" and a disease process, then experience shows that targeting proteins "upstream" and "downstream" from this protein will have therapeutic effects. In essence, all proteins in the cascade become validated drug targets, multiplying the number of potential targets some four- to ten-fold. This spectacular projected expansion is assuming many

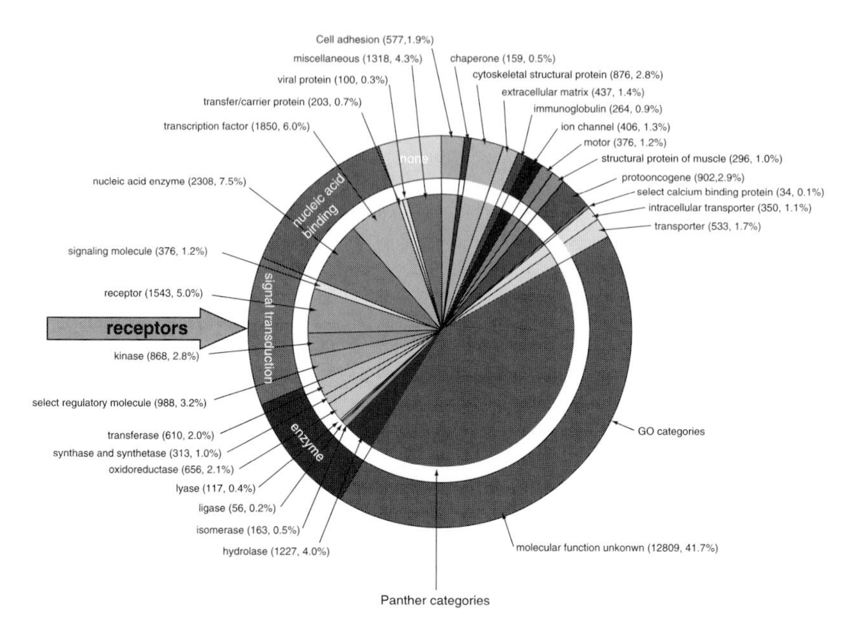

The complexity of target classes demonstrates graphically the complexity of the modern drug discovery process.

Figure 10.3 Human Gene Product Functions Suggest the Size of Target Classes

things: society's continued wish to have new drugs and pay for them; access to capital to develop these drugs; the spread of genotyping to lead drug development and later drug selection by the physician. It also assumes that the registration and approval process by the FDA and other authorities will keep pace.

The fundamental reasons whereby not all the newly identified targets will be used are: (1) they may be inaccessible to drugs because they are inside cells—intracellular—or buried deep inside intracellular organelles such as mitochondria; (2) they are not likely to be commercially interesting; or (3) there is a better target for the disease or diseases to which the target is linked. There will also be an improved distribution of classes of targets. The number of identified G-protein coupled receptors (GPCRs)—a popular target class in Pharma circles—will have grown from 75 to about 120, but will represent a reduced proportion of the drug targets from 30 to 20% because, hopefully, the industry will be working on other classes such as: RNA-protein, DNA-protein, and protein-protein interaction drugs. These interactions are getting more attention within basic research and are proving to be extremely important in regulating cell division during repair, healing, and cancer processes, and in regulating immune and allergic reactions. We know this from analysis of antibodies to proteins involved in these interactions or from research on transgenic animals where the proteins are altered in both structure and availability. However, we are not yet good at finding small molecules—new chemical entity drugs—to affect these interactions, although experience shows that a few breakthrough structures might lead to expansion in the industry. Such a breakthrough has recently been made by the discovery of *nutlin*, which controls protein-protein interactions regulating p53—a master protein of importance in cancer growth. Since the Pharma industry has no or very few earlier projects of this kind, their chemical libraries do not provide the necessary hits. Some great natural product-based medicines, for example, some antibiotics, do however work on some RNA-protein interactions.

TARGET CLASSES

What genomics has certainly done for the Pharma industry and drug discovery is it has intensified certain competitive races between companies. For example, despite genomics having revealed a more diverse group of potential targets, the race for the remaining GPCRs was probably the most intense such that all possible GPCRs have been identified. It may only be 3% of the genome, but it has 100% of the industry's attention. GPCRs were seen as the most rapid way to new drugs since the characteristics of the targets are very well known. While GPCRs control heart rate, blood pressure, acid secretion, and many vital processes, the largest number of GPCRs are unfortunately for the Pharma industry devoted to olfaction, and while sensation of scent and odor are important, our society seldom regards loss of smell as a serious disease.

The industry is still conservative and cautious in this regard. There is an expectation of a higher success rate with validated targets, especially target classes with which one is used to working. If one looks at the sales of medicines from a target point of view, one finds that the top 15 NCEs starkly demonstrate that the industry's experience with a target type is an enormously influential factor in the selection of clinical candidates and indeed in their development into drugs. Only 3 of the current[73] top-selling NCEs are "pioneer" drugs; the remaining 12 are "followers" or "me-toos." Looking at target types of these "top sellers," 5 of the 15 are GPCRs, 4 of the 15 are enzymes, 4 of the 15 are so-called "transporters," which carry

[73] As of mid-2004.

molecules across membranes, one is a "biological," and the remaining one is a receptor to estrogen in the cell nucleus. The companies are understandably rooted in their experience. They invented or copied someone else's drug—especially if it were an SSRI—copied their trial, and used their validated target, which was, appropriately, either a GPCR, enzyme, or the serotonin reuptake transporter. The industry is individually not that inventive, even if *in toto* the industry is collectively inventive. How many people in the world have developed a new antipsychotic from those thousands who have looked to do so?

If one looks at the six largest drug companies, the combined selection criteria (and the weight they carry in the decision processes) that determine the immediate strategy in new drug development is: a well-validated target (40%), a known class of proteins, that is, one that is likely to find hits in the chemical library (30%), a relatively easy high throughput screening assay (10%), solid structural—X-ray crystallographic—information of target with a bound ligand (10%), and, finally, that the target is an extracellular or cell-surface protein and easy to get to (10%).

Learning the Biology in Inflammation Leads to a Wealth of Targets

New ideas and a wealth of targets have to come from a deeper understanding of the biology of systems, not just from genomics and proteomics. A good example of old-fashioned "pathway finding biochemistry" comes from the investigation of inflammation, where one can find a wealth of targets. It's an important example because almost everyone has experienced inflammation, and it is a very common problem in many insults, injuries, and disorders that cause pain, whether a cut, a sprain, or rheumatoid arthritis.

The message from the complicated story is that after discovering all the biochemical pathways necessary to produce inflammation, the drug companies try to inhibit the process every step of the way. They stop the initial signal, the receptors that detect and transmit the signal, relaying of the signal by intracellular enzymes, and the production of the chemicals that cause the inflammation and secondary effects such as the pain. Inflammation is a natural and beneficial process, but when it gets out of control, long-term damage can result. But a word of caution, if you interfere with the inflammatory process there is a danger of interfering with many natural processes in normally functioning cells.

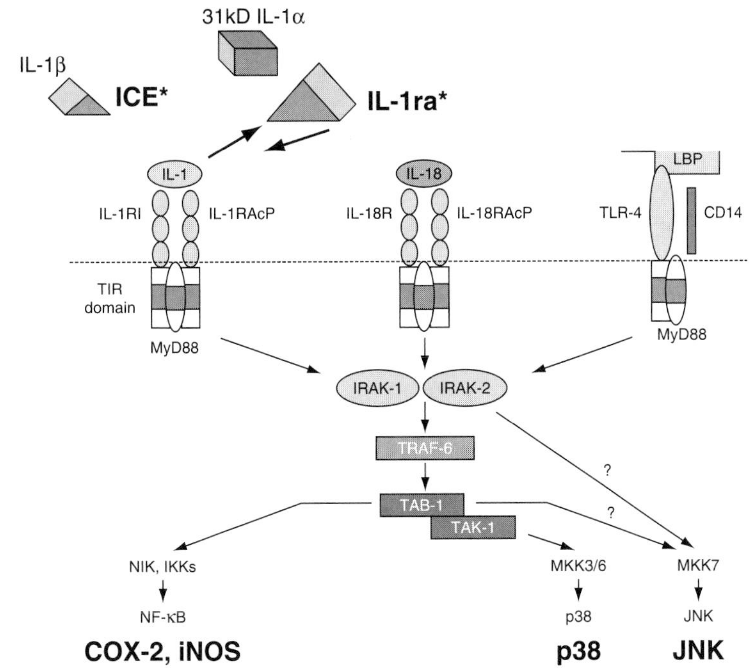

Schematic of targets in the biochemical pathway of inflammation. See text for explanatory notes.

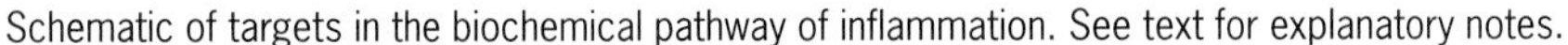

Figure 10.4 Targets for Anti-Inflammatory Drugs

Figure 10.4 shows a cartoon of the way we think certain inflammatory processes work. The illustration shows receptors in the membrane of a macrophage[74] in a joint that may be susceptible to arthritis, being stimulated by the naturally-occurring chemical ligands interleukin-1 (IL-1) and the bacterial lipopolysaccharide (LPS). The ligands attach to and interact with the receptors (IL-1R and LPS binding protein [LBP]) molecularly configured to recognize and receive them. This primary signal leads to inflammation through the production and release of prostaglandins. For example, one of the final biochemical pathways leads to an increase in cyclo-oxygenase enzyme 2 (COX-2) production, which is the target for the new COX-2 inhibitors, such as Celebrex, Vioxx, and others, used to treat pain. Other potential sites or targets for reducing inflammation would be: caspase inhibitors such as the interleukin converting enzyme (ICE); interleukin-1 receptor antagonists,

[74] A macrophage is a cell which when stimulated gives rise to the inflammatory response. It normally resides immobile in blood vessel walls but in response to, say, injury it becomes mobile and releases chemicals which in normal circumstances would attack foreign objects.

such as kinneret and inhibitors for the interleukin-1/2/4 receptor associated kinases (IRAK-1, IRAK-2, and IRAK-4), which are now in clinical trials. Similarly, you can have an effect by inhibiting natural inhibitory processes at the site of, for example, the mitogen-activated protein kinase (MAPK[75]p38), for which there are five compounds in trials for the treatment of rheumatoid arthritis (RA), psoriasis, and chronic obstructive pulmonary disease (COPD), and two for another kinase c-Jun N-terminal kinase (JNK), for which two compounds are in trial for stroke. Complicated? This is of course only half the story.

Cautionary Tales about Targets

The therapeutic area of inflammation provides additional examples of how the Pharma industry develops drugs based on an understanding of the biology, and how complex it can so easily become. It also begins to reveal how drug companies think, act and do. It's not always the science that drives the direction of discovery.

There is a relatively new, highly successful biological drug launched for the treatment of rheumatoid arthritis (RA). It was the only compound of a small company, Immunex. The drug is the **soluble tumor necrosis factor receptor**. Tumor necrosis factor (TNF) is a member of an important group of molecules called cytokines;[76] it is a highly potent promoter of the inflammatory response.

TNF is synthesized in macrophages—the mobile cells of the immune system seen above—where it is held as part of a complex called ProTNF (see Figure 10.5). TNF is released through the action of an enzyme (TNF convertase), and free TNF travels through the blood stream and binds to its trimeric[77] TNF receptor on a variety of cells, such as endothelial cells or neurons, which is the initiating signal of the inflammatory response that comprises a cascade of biochemical events. The very sound idea behind introducing *extra* free and soluble TNF receptors is that TNF will harmlessly bind to it rather than the TNF receptor in the cell membrane, and, thus, the inflammation signal is suppressed, or at least reduced. The soluble TNF receptor (etanercept) is marketed as Enbrel and, at the time

---

[75] Also known as microtubule associated protein kinases and early response kinases (ERKs). For more synonyms and even more complex descriptions of function, see www.brenda.uni-koeln.de/information/all_enzymes.php4?ecno = 2.7.1.37.MAPK

[76] Literally a molecule that "moves a cell"; practically, enzymes that act as intercellular mediators.

[77] Meaning comprising three distinct protein building blocks.

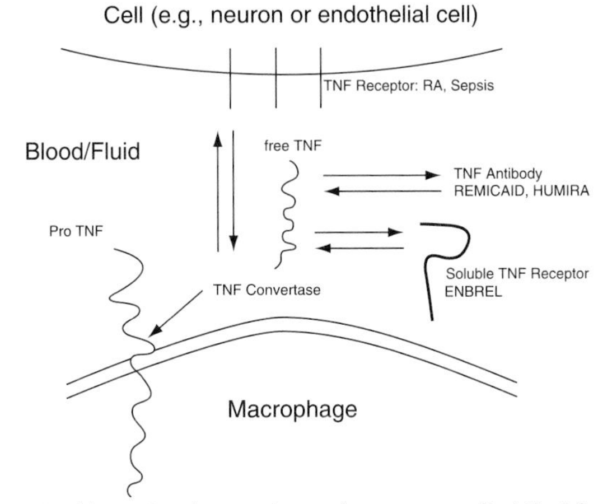

There are many ways to skin a cat and many places where one can affect the inflammatory response. Figure is of a "general" cell (above), say, a neuron or an endothelial cell, and a macrophage (below). See text for explanatory notes.

Figure 10.5 Proof of Principle for TNF Convertase Inhibitor

of this writing, had yearly sales in excess of $1.5 billion. It is the reason Amgen bought Immunex. Amgen shares Enbrel revenues with Wyeth, which markets the drug. It doesn't interfere with another major RA drug, methotrexate, so they can be used together. Enbrel is, by everybody's standard, a good thing. Rheumatoid arthritis is a truly debilitating disease, and it's a fantastic breakthrough. The only drawback is that being a biological protein it has to be injected. This makes it more difficult and expensive to administer.[78]

The fact that Enbrel works indicates that free TNF is indeed the villain in this system. Smart people in another company had the idea that one could obtain the same effect if you could inhibit the enzyme—"TNF convertase"—that is needed for TNF release. An enzyme inhibitor can be a small, orally administrable molecule.

As well as recognizing that one can decrease inflammation by targeting the source of the TNF as well as the free TNF, the company also recognized that this reaction was not just important in rheumatic arthritis, but also in

[78] Incidentally, a newer competitor, Remicaid, is only to be injected by doctors and provides a potentially useful supplement to doctors' income. This fact may not have been missed by the company's marketers. But an even more recent entry, the TNF–antibody Humira, can be injected subcutaneously—just as insulin is injected—by the patients themselves.

sepsis. Why is this important? Why the interest in sepsis? Because it is faster to do a trial in sepsis. Without treatment the patient is most regrettably likely to die within 16 hours. With partially effective treatment the patient will remain with severe bactoremia, or, with fully effective treatment, recovery is complete. The results will be known very quickly. A trial for rheumatoid arthritis takes many man-years. Your company may never want to market the drug in sepsis, but a successful trial in sepsis will give you the Proof of Principle for RA. Or will it?

The list of indications will be the same for all clinical candidates considered to interfere with TNF and its actions. The primary indication (i.e., the therapeutic area with the biggest market and greatest sales potential as well as suffering where the TNF response is excessive) is rheumatoid arthritis (RA). The secondary indication is inflammation, the tertiary psoriasis and sepsis, and it is also involved in muscle wasting and implicated in cancer.

The Proof of Principle is easy to show since you are likely only to have to show a reduction in blood-borne free soluble TNF. In addition, Eli Lilly has had a sepsis drug approved—in March 2002—and it works by lowering the cytokines interleukin-1 (IL-1) and TNF-α which again reinforces the Proof of Principle argument.

It is, however, not just a matter of Proof of Principle which governs decisions and the progress of projects in a pharmaceutical company. Here we have shown that there is a clear reason for pressing ahead with a TNF convertase inhibitor (TNFCI). Unfortunately, not all the decision makers at the drug company are so easily convinced.

The TNFCI, despite its clear, logical and appealing scientific origins and pedigree of good science, still hasn't reached Phase I. Drug discovery is a tough business.

chapter

11

SO MANY DRUGS, SO FEW ENTITIES

ALL ABOUT STRUCTURE

It will bear repeating, often, that *medicinal chemists are always the critical point in every preclinical development.* Who are they, and what so far have they come up with as medicines? The imagination and creative skills of the chemists may not have been fully exploited by the industry. This in itself gives hope for the future, but for the moment we have relatively few truly distinct structures, and we would like to exploit these.

While there is hope in that there are surely many more chemical structures to discover, the paucity of fundamentally different structures is a restricting legacy that still hampers the often-conservative industry.

The industry's medicinal chemists have over the years engineered relatively small structural changes in the basic structural chemical *scaffolds,* and both the clinical candidates and the backups were often from the same structural class. In order to arrive at fundamentally new scaffolds, there was a strong reliance on natural products. The main 8 to 10 scaffolds discovered in the beginning of the twentieth century are used heavily and represent the key structures of the majority of compounds of "historical" chemical libraries in all the major pharmaceutical companies. The answer to the question: "How many chemical entities are there?" may depend on one's definition, but almost all the chemical entities currently marketed can be traced to these scaffolds.

Box 11.1 The Ten Most Used Scaffolds in drugs before 2000[79]

The **hydantoin** scaffold as seen in the 50-plus-year-old, but still very much used, antiepileptic, phenytoin.

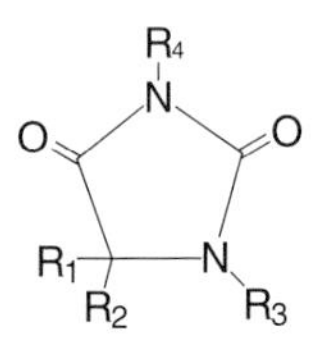

The **indole** scaffold as seen in the antipyretic or anti-inflammatory indomethacin and etodolac. (R3 and R4 also form a ring)

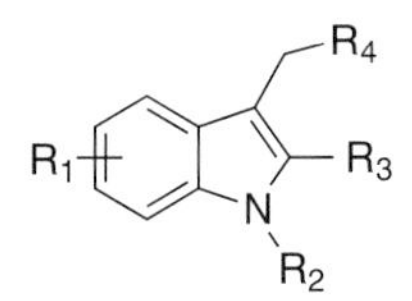

The *β*-**lactam** scaffold is the basis of the antibiotic penicillins and cephalosporins.

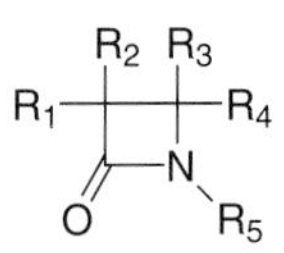

The **benzofuran** scaffold is found at the core of the antiarrhythmic amiodarone.

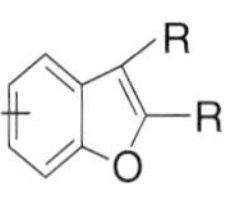

[79] Descriptions of most of these drugs are available in many online dictionaries, for example: http://www.online-medical-dictionary.org/ or http://www.nlm.nih.gov/medlineplus/druginformation.html

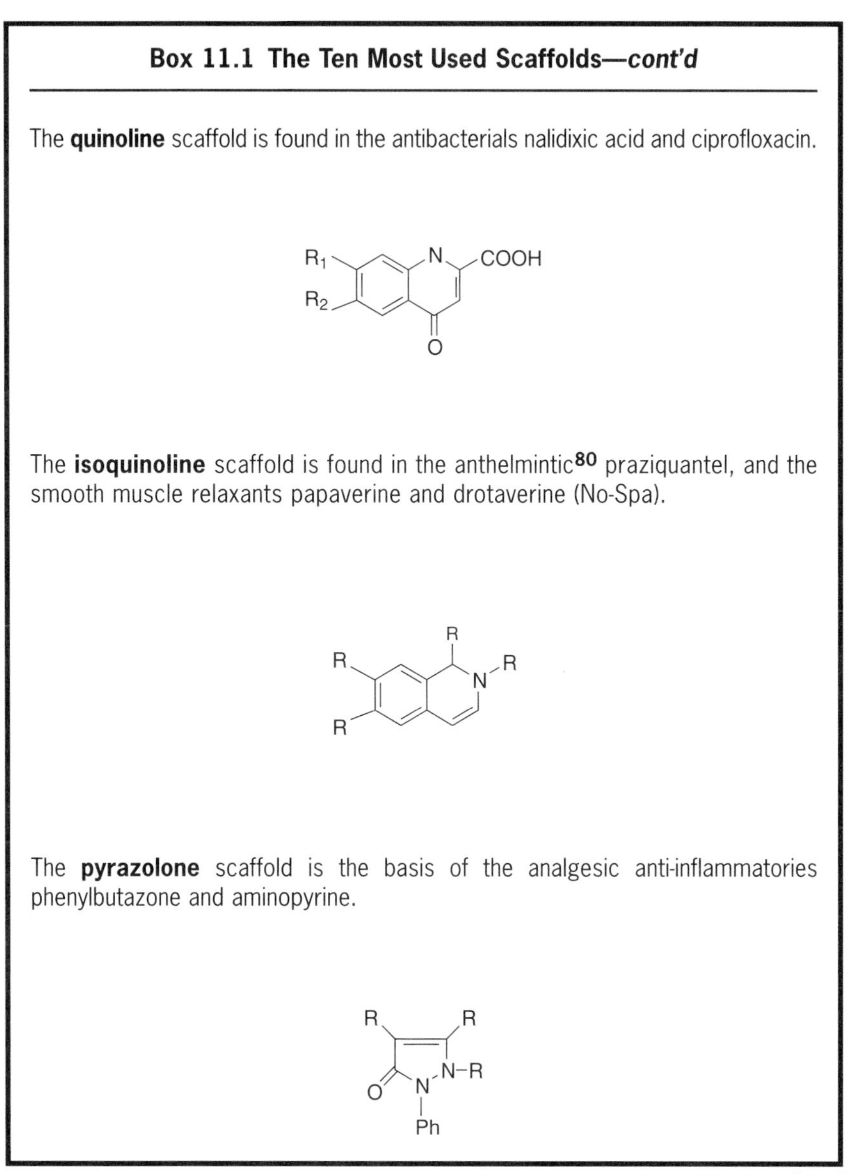

Box 11.1 The Ten Most Used Scaffolds—*cont'd*

The **quinoline** scaffold is found in the antibacterials nalidixic acid and ciprofloxacin.

The **isoquinoline** scaffold is found in the anthelmintic[80] praziquantel, and the smooth muscle relaxants papaverine and drotaverine (No-Spa).

The **pyrazolone** scaffold is the basis of the analgesic anti-inflammatories phenylbutazone and aminopyrine.

continued

[80] A medication capable of causing the evacuation of parasitic intestinal worms.

Box 11.1 The Ten Most Used Scaffolds—*cont'd*

The **pyrimidine** scaffold is found in the antivirals zidovudine, stavudine (anti-HIV), and sorivudine (antiherpes).

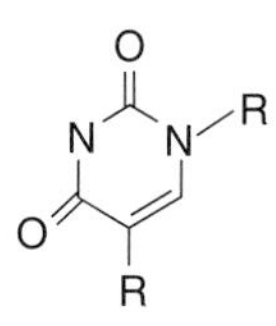

Other antivirals and antiparasitics are based on the **purine** scaffold. They include mercaptopurine (anthelmintic), acyclovir (antiviral), didanosine (anti-HIV), and vidarabine (antiherpes).

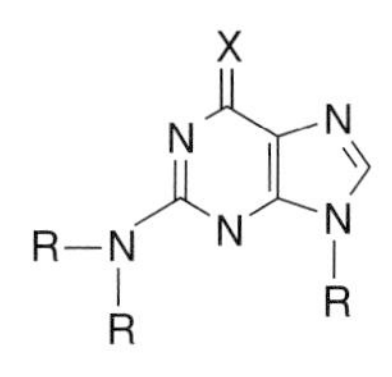

Finally, the scaffold behind one of the most famous class of drugs: the **benzodiazepines**, such as diazepam (Valium), and the other sedatives, hypnotics, and anticonvulsants flunitrazepam, midazolam, lorazepam, etc.

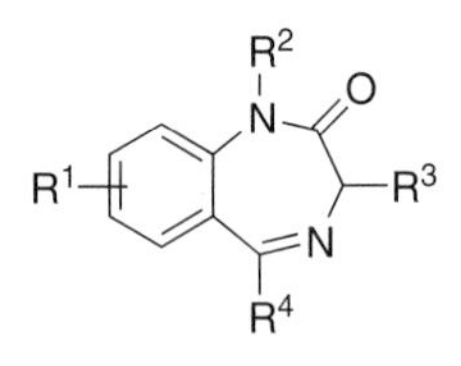

Figure 11.1 Leo Sternbach

Blockbusters

One of the famous fathers of modern drug discovery, Leo Sternbach, discovered and developed diazepam, which under its trade name Valium, became the first "blockbuster" drug with sales exceeding $1 billion per year. It is also accused of unleashing a culture of quasi-accepted drug use for semirecreational use in anxiety-ridden mainstream America. The problem isn't so much in the drug itself, but problems arose from its interaction with alcohol and through its induction and development of both tolerance and dependence.

Despite its side effects, it is still the "golden standard" for anxiolytic activity. Dr. Sternbach made such a good compound that this is the most robust and most fast acting anxiolytic, and efforts over the last 40 years have been to improve the therapeutic ratio by lowering the side-effects while retaining the efficacy. The many benzodiazepines are the mainstay anxiolytics. The modern antidepressants (Prozac, Paxil, etc.) were shown to act on anxiety, but these safer drugs are less robust than diazepam.

Leo Sternbach is also responsible for more than 100 patented compounds without benefit of high- throughput screening to assist him. Perhaps his main legacy is that he created the dream of all drug companies with his blockbuster. They want a few products that are easy to market, and they should all sell a lot.

Making Sense of the Chemistry

Who does what in the chemical laboratories of Big Pharma? The chemical library comprises many parts. The core is the historical library, the chemicals that the company has been making over the past 100 years from dyes onwards. The majority of these compounds reflect the historical success in particular therapeutic areas, which may be, for example, cardiovascular, analgesic, or antifungal compounds. The second group comes from the combinatorial chemists who can make molecules "on demand," just in case. They take the basic structure of the molecule in question and systematically substitute atoms or molecular groups to modify the molecule. Instead of, say, a hydrogen atom at every point on the structure or scaffold, they can add methyl (CH_3), ethyl (C_2H_5) up to, say, $(CH_2)_7$ open chain—"aliphatic"—groups. Instead of a CH_3 they can try a silico-fluoro group (SiF_3). Some molecular substitutions would be less likely to be biologically useful (e.g., cobalt for carbon) and will be avoided. A systematic expansion of existing structures is made, and a large number of compounds is generated using just a few sets of building blocks in thousands and thousands of combinations, to create the combinatorial libraries of 100,000 to 1,000,000 compounds.

A company needs to go beyond this and contracts combinatorial chemistry companies to construct additional combinatorial libraries. They may ask a company to create a library from a combination of "building blocks." One chooses a large set of blocks, say five to six different sets each with dozens of different substituents and makes the final compounds by connecting five to eight of these complex building blocks together in different ways. If you start with sufficiently complex building blocks, there usually are a sufficient number of combinations for the molecules synthesized to be unique, that is, if selected or chosen as a drug it stands a good chance of being patentable. Obviously, different chemical companies can request the same efforts from different providers, and the companies can end up synthesizing identical molecules. If this turns out to be a valid clinical candidate, then the winner is the one that applied for the patent first. It has happened often as medicinal chemists of different companies are trained in similar or identical academic laboratories, and choose similar strategies when optimizing a molecule to fit a very narrow well-defined space in the drug-target protein molecule.

The library can be vastly expanded because of a technical breakthrough. One no longer needs massive quantities of chemicals to test. With *in silico* screening one can test the theoretical fit of the compound to the target using three-dimensional computer models of both structures.

Unfortunately for many of the industry's favorite targets (GPCRs, ion channels, etc.) no structures exist yet. Good computer programs can go through millions of putative drug structures, some of them only existing in the mind of the computational chemist; in other words, they have not yet been synthesized. If a molecular structure shows promise in the computer modeling, then it will be synthesized and a whole little library around it will also be made. The compounds can be made in small quantities as the assay volume has shrank from milliliter (10^{-3} liter) to the nanoliter level (10^{-9}), or in some cases, picoliter level (10^{-12} liter). With biotechnology one can also make a lot of the target proteins by expressing these in bacteria or yeast in the same way most biologicals—protein drugs—are made. With robotic technology one can synthesize millions of compounds and screen them at the rate of 1 million compounds a day on a given target.

Where does man insert himself back into the process? Once the "hit" is detected, the molecule has to be checked for stability, acute toxicity, and possible drug interactions with commonly used other drugs. Various structural changes may then be introduced by some of the 1,000 medicinal chemists at the company. The pharmacokinetics and toxicology are then determined to create the clinical candidate, and, if one sensibly plans ahead, an alternative as backup is developed in parallel.

The fact that medicinal chemists with special skills are crucial components of the drug discovery process is evinced by the observation that very many of the best chemists have come from relatively few mentors in even fewer laboratories, such as the ETH in Zurich, Switzerland, Oxford, UK, Columbia University and Sloan Kettering in New York, and Harvard and Pittsburgh. Considering how large an economical activity the Pharma industry is there are only a handful schools where medicinal chemists are trained—as opposed to synthetic organic chemists, which are trained by every university with a science faculty and a chemistry department.

chapter

12

HOW TO FIND A CANDIDATE DRUG

SELECTING CLINICAL CANDIDATES FOR TRIAL

Once a company has selected its clinical candidate, it has to be made in bulk—in hundreds of kilogram quantities, which itself is not trivial, to prepare for entering clinical trials. It is important to understand the significance of the sequence of phases in clinical trials. The stages are:

Phase I
The stage of drug development in which drug safety is determined in a small group of 20–40 healthy volunteers; cost: $1 million

Phase II (A & B)
The stage in which drug effectiveness is determined in a small group of 100–300 patients; cost: $20–100 million

Phase III
The stage in which large-scale safety and effectiveness are ascertained is as many as 20,000 patients; cost: $200–300 million

Phase IV
The later nonobligatory stages where approved drugs are tested against new indications for an extension of the prescription labeling

A basic maxim in drug discovery for Biotech and Small Pharma companies is that no matter how great a scientific idea is behind a clinical candidate, **if you cannot bring a clinical candidate to a Phase II trial you actually will spend more money on discovering your drug candidate than what you can make on selling it to a larger company**. Small companies that can get to Phase I, or even early Phase II (Phase IIA), will just about break even on their investment in today's market for drug development. Should a small company take the risk of failing in Phase I? If it wants to realize the value of its research then YES, but it should have more than one Phase I compound.

Why would a scientist at a start-up or a small company be looking to "sell" their candidate drug? The reason is that it costs an awful lot to take a candidate further. It is expensive to carry out large-scale efficacy trials (Phase III) in 20,000 people for two to five years or so, depending on what long-term benefit is to be proven.

It is worth reiterating that the decision to select a clinical candidate is not purely a scientific decision about the estimated human efficacy of the drug; it is also about whether one can prove efficacy and near absolute safety, and prove it cost effectively. There are many smart people in pharmaceutical companies who have to make the decisions. Alongside the R&D scientists who have often come up with the idea of a drug target, validated it, made the appropriate drug candidate, and tested it in animal models of the disease are the even ***more*** influential clinical trial people and the marketers or "marketeers." Unfortunately, it might be argued, if there were any doubts about the candidate, or making money from it, then the doubters would prevail, if only to allow resources to be focused on projects where there is unanimous support. Interestingly, one of the most valuable resources for the companies is the medicinal chemists. The medicinal chemists remain the critical point and bottleneck in every preclinical development, except for the development of the much fewer biologicals, therapeutic antibodies, and recombinant proteins. Circumstantial evidence for this comes from the fact that after a merger the most secure job is that of the medicinal chemist. The preclinical biological scientist is the least influential in the decision making.

The Ideal Clinical Candidate According to the Company

The industry average is that only 1 of 10 clinical candidates selected for entry into Phase I or entry into man (EIM) will reach the stage of approved drug that can be marketed. Some will fail early at low costs, but others may draw $300 million before the company knows that a drug did not make it.

"The best clinical candidate is a candidate which has been already marketed." That is to say, the company will market their successful drug for another indication. "We have already tested it in a million people. It is safe. It is ours." All they have to do, for example, is make a new preparation as a cream that absorbs faster, and if this new formulation can be patented, the lifetime of the drug is greatly enhanced. Today, "life-cycle management" of existing drugs is a bigger thing than discovery of new medicines as companies come to "play it safe." That's the perfect clinical candidate because the risks are minimal. If a company wants to market a drug for new indication, the safety is usually known and the efficacy has to be ascertained in clinical trials. But there is usually a lot of confidence in establishing efficacy as the whole effort is usually prompted by the clinical observation of efficacy in the "new" indication found and reported during "off-label" use.

In the absence of a history of an already marketed drug, a company has a long list of requirements for any clinical candidate (see Table 12.1). These are very important indicators to the company in deciding whether to go ahead, put in the $30–300 million, and wait out the four to five years of clinical development. Preferably there should be clinical proof that it works, or at least strong evidence of efficacy in an appropriate animal

Table 12.1 Favorable Characteristics of Clinical Candidates

The clinical candidate should preferably have all the following attributes or characteristics:

- Act on a validated clinical target
- Be efficacious in relevant animal models
- Be "clean" in genotoxicity tests
- Be "clean" in acute and chronic toxicity tests
- Have favorable pharmacokinetics in dose and form
- Be easy to synthesize with reasonable "cost of goods," and good availability of starting material
- Lack drug interactions and preferably ***not*** be metabolized by cytochrome $P_{450}3A4$
- Have a reasonable "backup compound"
- Be a candidate for prescribing by general practitioners rather than specialists
- Have a competitive advantage, preferably by being first in class, and accelerated FDA review
- Have a robust patent with long life
- Have minimal direct competition: drugs on the same target
- Have minimal indirect competition: drugs for the same effect

model of the disease. This is of course providing such animal models exist. They do in infectious disease, oncology, and endocrinology to a much greater extent than, for example, in psychiatry. We really have no rat model of hallucinations that is reliable.

It has to be toxicologically "clean"; neither general nor genetic toxicology would be tolerated when delivered initially ("acutely") or over the long term ("chronically").[81] It has to have a favorable drug profile in that it must act in doses suitably small and be easily delivered. In other words, a daily or weekly tablet is better than a twice-daily injection; it would sell better. In addition, it should be reasonably inexpensive to make with easily available starting materials. It should be one likely to have minimal interactions with other drugs. The most reliable way of ensuring this is for it not to be metabolized by the most common enzyme for drug metabolism, the mixed function oxygenase: cytochrome P_{450} 3A4.

Why is this important? Each and every one of the drugs that were withdrawn in the period 1998–2002—and there were seven of them selling for a collective $11 billion per year—failed because their metabolism went through this enzyme, where they competed biologically with 70% of the drugs on the market today. If your new drug is being metabolized in this biochemical pathway, it will suffer from biological competition. All assumptions about dosing and the like are compromised if one cannot make sure it is not going to be prescribed at least in some patients taking some of these other drugs.[82] If drug interactions are indicated, the FDA is more likely to reject the new candidate drug if only because the others have already proven benefits and safety, and why would competitors agree to take them off the market when they sell well? To make predictions more difficult about drug interactions we need to remember that the rate of drug metabolism may vary by as much as tenfold between healthy individuals and patients.

Of course, however it is metabolized, the resultant *metabolite* of a drug should be inactive and preferably benignly excreted or secreted.

Before a compound is put on the expensive path of sequential clinical trials, a prudent company insists on the chemists having a backup

[81] As mentioned before, there may be some tolerance in the case of oncology drug candidates because of the devastation and poor prognosis of the disease and because the drugs themselves are often intentionally toxic to the cancer cells.

[82] It is extraordinary, but not surprising, that most drugs are metabolized through this pathway. It is not surprising because virtually everything that is lipophilic enough to go through cell membranes to hit an intracellular target will bind to P_{450}3A4. However, many other pathways should be available to chemists to avoid this pitfall. There are 12 other iso-enzymes of P_{450} and drug metabolism can go through pathways of glucoronylation, to glutathione-conjugation; a million ways to avoid drug interactions!

compound. This is hard for some scientists in academia, Biotech, or Big Pharma to reconcile with their drive for biological innovation. Academic scientists can recognize a new target, develop a candidate drug for the target, and publish the results in journals like *Science* and *Nature*. There after, in trials, the company would find something wrong with it—be it poor efficacy, excessive side effects, or unexpected drug–drug interactions—and the quest would finish there. By that time, three other companies have made a better compound for the target, which, thanks to you, they have now read about. The companies that produce new drugs have backup compounds. The companies that start without backup compounds do not often produce new drugs. Simply put, if you believe in your target you should develop more than one compound to hit it. Many times there are several indications that involve the particular target and it may make it possible to sell both the front-runner and the backup—for different indications—if they both successfully navigate through the clinical trials.

If the drug is going to be widely prescribed it is better if it is a drug that is going to be prescribed by general practitioners (GPs) rather than specialists. This, in itself, means that it is likely to be prescribed alongside other drugs with which it has not been rigorously tested for interactions. But the potential financial advantage of having many more physicians prescribing your drug is very attractive. For example, the major market for antidepressants goes through GPs, not through psychiatrists. In particular, if you can show that the antidepressant works as an anxiolytic—as was proven for the SSRIs—then more GPs will prescribe it. Of course, it is a lot more expensive to market to GPs than specialists where there may be as few as 350 neurologists[83] in countries such as the UK.[84] In reality, a company would want a balanced "portfolio" of drugs for over-the-counter (OTC) sales to be prescribed by GPs and then specialists.

Another critical piece in this puzzle of clinical candidate decision making is whether the drug has a good competitive advantage against existing treatment, or potentially even better, if it is "first in class." Apart from the obvious reason that it is usually—but not always—better to be first and be established in the market first, there is the other reason that most overworked physicians—with medical school decades behind

[83] This is a small number—for a country of a population around 60 million, that is, 1 per 177,000 according to the Association of British Neurologists. France has roughly 1 neurologist per 39,000 of its 60 million residents. In the United States there are 18,000 members of the American Academy of Neurology for a population five times the size (i.e., 1 per 17,000).

[84] Note that in the immediate past, drug companies used to be able to entertain physicians and their spouses, for example, while attracting their attention to any new product. Now this is much more highly controlled, especially in the United States.

them—usually only remember about three drugs in a class for each indication. It is not necessary for them to remember more. It is likely if the first three drugs in a class have not worked that the next three or four in the class are also likely to be ineffective or produce unacceptable side effects in their patient. One must question whether the world actually needed seven SSRIs to treat depression by the same mechanism when 30% of the cases are treatment resistant to SSRIs and are waiting for a drug with another mechanism or aimed at another target that may help them. But playing it safe makes the seventh SSRI and the fifth COX-2 inhibitor more attractive as a "safe bet" for a big enough company with marketing muscle, much safer than a fully new drug.

The value of being "first in class" has been often reduced in recent years. The competitive advantage afforded by being first is *very much overrated*. Being first in its class does still enable one to receive an accelerated review from FDA. But only 4 of the top 20 sellers from 1985–2000 were considered in review as a "significant improvement" over existing therapies. All the others were "just like the others," the so-called me-toos. As we have mentioned earlier, some of the successful chemical innovations were not very significant scientifically since they are derivatives—or "copies" with enough to distinguish them so patents could be obtained—of "first-found structures." Of course, the expected sales life of the drugs entering an established class have the effective patent life of the original in that once the patents of the first in class expire, generic prescription and, where appropriate, OTC drugs would compete with all of the drugs in the class.

A good recent example that may be of potential historical interest is the Viagra versus Levitra and Cialis story. Viagra, the first drug successfully to treat erectile dysfunction in a large population without complicated injections, established the market by identifying a market need that had hardly been recognized or admitted in public (i.e., outside the physician's office) and even there it was seldom brought forth.[85] Levitra and Cialis, which are chemically very similar,[86] could enter the market rapidly on the heels of Viagra with a prospect of even greater potential in a market established by the innovative drug. Now Viagra, Levitra, and the most recent Cialis are being promoted to treat "female impotence," as it has been coined, or more correctly "female sexual dysfunction," even though

[85] Viagra is a phosphodiesterase-5 enzyme inhibitor. It was not made to treat impotence—that was a side effect—almost not discovered because men in Phase II did not want to talk about unwanted erections.

[86] A tighter patent for Viagra may have stopped Levitra from being able to be launched.

physicians are very divided as to whether, for example, lack of sustained interest in the long-term partner should be "medicalized" as a malfunction needing intervention. Of course, now that these drugs are on the market, it is likely that women will determine if they add to the quality of their health or life experience. The drug companies, Pfizer, Eli Lilly/ICOS, and Bayer/GlaxoSmithKline, may of course sponsor research and invite debate and discussion, but they cannot say that their product "relieves female impotence" without real data.

In the selection of clinical candidates to pursue, the company does take into account competitors in the traditional sense of those that would compete for sales and reduce the profit potential. Some drugs are dropped from development plans because they are expected to be poor competitors with established drugs, especially where the competitor has a more effective sales force in the therapeutic area.

It is not only direct competitors that affect the same target in a similar way that one has to be concerned about. If one is working on attacking, say, a particular tumor-specific target, then competition can be ***indirect***. A tumor needs its own blood supply to grow, and blood vessels develop throughout a tumor in a process known as ***angiogenesis***. Without this new blood supply, tumors would not grow, and if a competitor comes out with a truly good angiogenesis inhibitor to strangle tumor growth, the bottom would fall out of your specific tumor-cell inhibitor market. This "**indirect competition**" is extremely important and almost always underestimated. Every Biotech tells an enormous story about the fantastic and specific protein drug target they found. This may indeed be true; but it may not be the only way to reach the same therapeutic effect. Many times a new drug will become part of a "cocktail" consisting of the drugs in the category of indirect competition to treat the disease. Physicians and patients buy an improved therapeutic effect or increased convenience of administration, such as pill vs. injection vs. infusion. They don't buy a new mechanism; they couldn't care less. Every drug marketer will explain this to you!

The Proof of Principle Principle

Inventive evaluation of existing drugs' performance has led to innovations in drug design. The side effects of early antidepressants—the tricyclics (see Box 12.1)—were at least partially due to their general inhibitory effect on acetylcholine (ACh) transmission. This is not surprising given that the prevailing hypothesis was that antidepressants' action was partly attributed to their anti–ACh effects.

The idea to make serotonin uptake blockers came from the desire to make an antidepressant without anti-acetylcholinergic side effects, such as dryness of mouth,[87] even though the prevailing dogma was that the acetylcholinergic effect in the brain was part of the antidepressant effect achieved mainly by blockade of noradrenaline and serotonin uptake. How can such a dogma be dismissed? First by making a compound that doesn't have any anti-ACh effect and then testing it. Did this new class of drugs still work as antidepressants? Indeed they did, and they were superior in terms of side effects, and, in some cases, in terms of therapeutic effects too. The class of one of the most sold and medically most successful drugs was born: the SSRIs.

This was a **Proof of Principle** trial for Zimelidine. Zimelidine was the first in class drug that was withdrawn once already on the market and taken by 30,000 people. It was replaced by the widely successful fluoxetine (Prozac), a "me-too follower." The Zimelidine trial showed that a drug that only inhibits serotonin uptake, not uptake of noradrenaline,[88] and that does not have or cause an anticholinergic effect, is an effective antidepressant.

The story does not end here. Later Phase IV studies showed the SSRIs to be effective in treating social phobia, and postmenopausal symptoms, and this expanded their use far beyond the use of the tricyclic antidepressants the SSRIs were designed to replace. The tricyclics were—and are—good, safe, and efficacious drugs, but they had—and have—the dry mouth side effect. This is not very dangerous, but it is a very unpleasant side effect.

A company has to spend about $20 million in preclinical research and development to make a drug that is safe enough to put into enough people to find out whether or not that part of the pharmacological effect is essential. Forty percent of all human trials are Proof of Principle trials, and the drug company spends $20–40,000,000 just to find out whether they are on the right path. Biotech companies don't think of this since they are focused on *their* molecule. Big Pharma often know that the particular molecule/compound they are testing is not going to make it as a marketed drug, but is good enough and selective enough to tell them in human trials that the mechanism and the target it affects are worth pursuing.

---

[87] The salivary glands of mammals are controlled by ACh, and the drug that affects brain ACh is almost certain to affect the salivary glands and make the mouth dry. If the patient is anxious, dry mouth can make him or her very noncompliant. It is not a minor side effect for these patients.

[88] As adrenaline is called epinephrine, noradrenaline is called norepinephrine in the United States and other countries.

Box 12.1 Structures of Representative Classes of Antidepressant Drugs

The structures of the first four compounds below have all been used as antidepressants. They all "work," which really does count, but not without difficulties. Iproniazid, a monoamine oxidase (MAO) inhibitor, turned out to be too toxic to be of continued use. It, along with all MAO inhibitors, has the problem of necessitating dietary restrictions (no red wine or cheese). The next revolutionary class of antidepressants was the tricyclics. Imipramine, though, has a 30% dropout rate because of its dry mouth side effect. This is an opportunity for a drug company. Efficacy may be good, but if a drug's side effects are so bad that they affect performance, there will be an opportunity. Enter the SSRI fluoxetine (Prozac), which is much better accepted and is very widely prescribed, but it takes 14–20 days to work (just like the tricyclics), and it inhibits sexual drive in many patients causing some anhedonia[89] on its own. This is not good

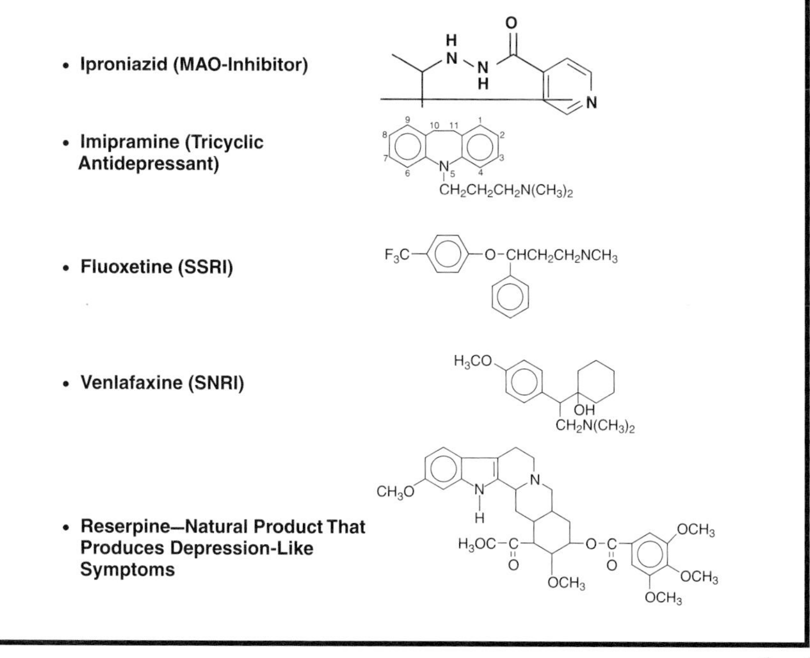

continued

[89] The inability to gain pleasure from normally pleasurable experiences was a concept first identified in the 1890s, then ignored before being more recognized from the 1980s as a major symptom of depression and other disorders. There are now animal models of anhedonia reported. (See http://www.biopsychiatry.com/anhedonia.html)

Box 12.1 Structures of Representative Classes of Antidepressant Drugs—*cont'd*

for an antidepressant. Thus, there is a new opportunity for the mixed—serotonin and noradrenaline[90]—reuptake inhibitors (SNRIs) like venlafaxine. However, even venlafaxine takes about 14 days to act, and although it is regarded as more robust in its antidepressant effect, the race is on to find a mechanism that will give a faster antidepressant action. Consider what an advantage it would have to cut the suicide rate among depressed patients. Proof of Principle on monoamine effects in depression came later, with lower levels of the monoamine serotonin (and 5-HIAA) being found in the brains of suicide victims. Scientists are looking for similar robust biomarkers in depression. Faster onset of antidepressant action would make it almost unethical not to start with the fastest acting drug because of suicide risk. A natural product like reserpine actually induces depression symptoms, and works by depleting both serotonin and noradrenaline stores. This finding served as a further Proof of Principle for the development of tricyclics, SSRIs, and SNRIs. (See text for more details.)

Picking the Right Thing to Measure in a Trial: Surrogate Endpoints or Markers

From clinical trial physicians' and biostatisticians' points of view, *Proof of Principle* is paramount. Then it gets difficult. One has to predict the effective dose, and one doesn't have too many opportunities for error. The task does not get easier as the highest and lowest rate of metabolism between individuals may vary ten-fold! This is why physicians often have to adjust the doses. It can be a significant problem and many drugs fail for either being tested at too low a dose and found ineffective, or too high a dose and having limiting side effects. This is not surprising to chemists.[91]

---

[90] In the United States the preferred names are norepinephrine and serotonin, and in Europe the preferred terms for the identical molecules are noradrenaline and 5-hydroxytryptamine or 5-HT.

[91] Failing to determine the right dose may kill a drug. Big Pharma projects often have 10 to 20% of their budget on making the drug in a form that can be followed by positron emission tomography (PET) or other imaging in the human body to tell how many of the possible sites of action the drug reached at a given dose. Biotech almost never has these resources, nor does it recruit or commission them.

With some chronic slowly progressing diseases like osteoporosis, rheumatic arthritis, and Alzheimer disease, it can take a long time to find out whether the desired therapeutic effect is achieved and maintained over years of treatment, and to discover that taking the drug slows disease progression or ideally stops it. This is why the designers of the clinical trials are always looking for "**surrogate endpoints**"; that is, you look for something which indicates the therapeutic effect indirectly. Drug companies do not have the time to wait for the actual therapeutic effect to manifest itself. It is much better to find out that the drug doesn't work early in Phase I or Phase II. What would be a good surrogate effect? Well, if you had a drug that is aimed at rheumatoid arthritis (RA), then the anti-inflammatory drug would have to be tested for six months to two years. But it is known that prostaglandin is a proinflammatory substance; that is, the blood levels of prostaglandins increase during inflammation. If your anti-inflammatory drug reduces the level of prostaglandins—that is, reduces prostaglandin synthesis—under conditions where inflammation is going to occur, then you have a Proof of Principle that you can detect after only six hours. In addition the surrogate marker in volunteers helps you to establish dosage levels for a long-term trial against RA directly.

The only time that surrogate markers are of no value is in the discovery of antibiotics and antivirals, where with PCR to amplify the bacterial or viral nucleic acids within a few hours, one knows directly whether the antibiotic or antiviral drug works: it kills microbes or stops their division, or it does not work. Hence, antibiotics almost never fail on efficacy, but their nasty side effects may yet stop for them or limit the dose patients tolerate; therefore, they might not show efficacy because the dose cannot be achieved that would kill the microbes.

Ideally, one would want to be able to test the drug alone against placebo. If the disease is serious or life–threatening and there is a presently available therapy with which all patients diagnosed with the disease are treated, then it is not ethical to take them off their drug that works for the chance that a new one may also work or work better. A typical example is found in antiepileptic medications where neurologists would hesitate to cease prescribing an effective existing antiepileptic and risk convulsions,[92]

[92] As well as the seizures themselves potentially causing brain damage and worsening the condition, epilepsy carries the additional risk that convulsions can occur anywhere at any time and a patient may be at physical risk. A neurologist might switch antiepileptic treatment if the seizures were not adequately controlled by a drug, or if the side effects (e.g. weight gain, decreased alertness) were excessive or intolerable for a particular patient, for example, if they were wishing to become pregnant. Some epilepsies are hard to control and are "refractory."

or in Parkinson's drugs, where taking patients off their medications will cause more tremor, general rigidity, and so on. In some cases, the FDA and other authorities demand to know that your drug is better or at least equal to existing treatments. A company must show that directly.

Sometimes testing alone can be done by waiting for newly diagnosed patients to whom to give the new drug, while watching closely for efficacy. If the new drug is not working, then a rescue dose of the proven old drug is administered. Sometimes one does a "crossover design" trial where patients are on a known drug for a week and on the new drug for a week, so the effectiveness can be compared within the same patient, who is, importantly, never without medication. Neither the patient nor the physician should know on which drug, old or new, the patient is at any given time. This is an example of a so-called "double-blind study."

Sometimes your drug is not going to be a stand-alone medication. It is from the beginning designed to enhance or complement the effects of another drug. No drug company really wants to design such drugs as first choice, but it happens since in complex diseases a single drug often does not remedy all symptoms. A typical Parkinson patient is on two to three drugs as the disease progresses.

If, for example, you are trying to make a drug against multiple sclerosis (MS), a slowly progressing disease with many recurrences and relapses for which there happens to be an approved treatment, β-interferon, then you have to combine and compare it against β-interferon. If you are a Big Pharma and your drug reaches the market as an add-on treatment with β-interferon, you are now dependent on the marketing machine of Biogen or Schering. You would either have to market for Biogen or buy the company. That may be a better solution in the long term because it's cheaper. Biogen was in fact bought in 2003 by IDEC Pharmaceuticals, not yet a Big Pharma, and the β-interferons became a good example of companies competing on side effect profile and on price. The much larger Schering-Plough and Aventis were successfully gaining market share from Biogen, and the sale to IDEC was a proper defensive response to this.

Finding Clinical Candidates in Your Portfolio

If the medicinal chemists were successful their compound in the chosen dose will hit only one target. Still-unwanted side effects can often be seen in the cleanest, most selective candidate drugs or medicines. Nonspecificity of a drug comes from its target occurring in more than one site.

One can turn this nonspecificity into an advantage and have therapeutic efficacy in more than one syndrome.

Clinical candidates can come from careful observation, serendipity, or luck, or all three. One real example comes from judicious observation of results from new technologies. New genomic-based techniques using *microarrays* hold real promise in drug discovery. Simply put, all genes or, rather, their primary gene products, are organized in a reproducible way on a chip in an array of up to 60,000 micro-wells. These microarrays are commercially available or can be made in a company's own laboratories. By preparing tissues from, say, biopsies in specified ways and cross reacting the tissue with the microarray, one can detect which genes in the sample were active—upregulated—or inactive—downregulated—compared with a standard sample. In this way scientists hope to determine which genes are over- or underactive in very many disease states.

In this example, a scientist at a Big Pharma noticed that a new sodium channel was upregulated in spinal cord samples in an animal model of pain. She isolated the ion channel and made, by a process called transfection, a cell-line that expressed this sodium ion channel. She could then examine the characteristics of the channel more closely and found that it was blocked by one of the drugs already made by the company. Perhaps not surprisingly, since most sodium channels are associated with excitability in nerve cells, the drug was an antiepileptic—a drug that blocks excitability in nerve cells. The same sodium channel blocker also turned out to have excellent antidepressant properties in bipolar (manic-depressive) patients.

This finding is not that surprising to scientists in drug discovery. The hypothesis from this is that since the sodium channel was found specifically in tissue associated with the experience of pain, perhaps the antiepileptic would be effective against pain. The most attractive thing after finding the target, and possible clinical candidate, is that one doesn't have to perform a new clinical trial to have a Proof of Principle to see whether the drug is effective against pain. You may get away with a meta-analysis. The drug is already being used in patients who see neurologists, so all one has to do is ask the neurologist to conduct a survey among the patients and ask: "Did you ever experience pain relief on this drug?" If some of the patients say: "Come to think of it, I went to the dentist and it didn't hurt that much," you have a Proof of Principle in man. This may not be the greatest scientific discovery, but it is terribly interesting. This kind of anecdotal evidence in most cases makes the company conduct a straight-out trial to seek official approval for the drug in the new indication as well. This process expands the life cycle of a safe drug and brings relief to patients in a new indication. Many times this is the only way we get new drugs into underserved indications. Pricing and dosing for the "new" indication are, of course, big discussion points within the company.

In this case the chemists put two polar groups on the drug, so that it doesn't enter the brain, where it might be sedating, but remains peripherally active against pain. The drug company then had an outside company test it against around 63 other ion channels to show it was selective against this particular sodium channel, before conducting a relatively fast clinical trial for postoperative pain, which only lasts four days.[93] If it works against postoperative pain, then the company applies to the FDA to have it approved as an analgesic agent.[94]

Fortunately the toxicology is "clean" for this slightly modified molecule, which, incidentally, is easy to synthesize with only six synthetic steps. The derivative structure and several other structures that were potential backups to the original antiepileptic were covered by the original patent application (i.e., the company already owns this NCE), but pain was not listed as an indication. This means that the company can now apply for a ***use patent*** for this chemical innovation. The target is old, the drug is old, but the label is new. This is a formula borrowed from car companies.

Targets for the Optimists

Targets are not seen in the same light by all participants in drug discovery. A clinician might take a simple but extreme view when looking at, for example, diabetes. Insulin controls the level of blood glucose, and when

---

[93] Incidentally, Biotech, because of inexperience and limited cash resources, often forgets to outsource selected projects and tries to do everything in house. On the occasions when it finally does, the drug candidate has already consumed large resources and there is reluctance to ditch it based solely on data indicating lack of selectivity. Big Pharma conducts these tests early on and in general throws out drug candidates more efficiently than Biotech does.

[94] Of course, it might not be that simple in practice. This particular candidate caused—admittedly at 50 times the suggested dose—a prolongation of the so-called QT wave in the electrocardiogram (EKG). For this the FDA requires testing against a particular potassium channel—the [HERG K^+] ion channel—that is known to be the major reason for QT prolongation. The problem with QT prolongation is that it may unmask a hidden arrhythmia of the heart and thus may cause very serious cardiovascular side effects. Again Big Pharmas test compounds for this very early on even if it is expensive and a "slow throughput test."

insulin levels are maladjusted or out of control, severe problems of diabetes are exposed. Therefore, it might be argued insulin, and the level of insulin, is the real validated clinical target that therapies should aim at controlling.

A molecular biologist would argue that insulin is not the only target but that insulin receptors—through which insulin has its biological effect—are the real targets. The counterargument to this runs that unless you know the biochemistry for sure[95] the receptor as a defined target will always be in doubt; the level of insulin remains, arguably, the true target and the easiest marker to measure for controlling the disease. Patients do it themselves.

Undaunted, the molecular biologists, who have founded many Biotech companies, argue that once you have identified a *transcript* or a *gene product* coded by it that is specifically expressed in humans or in some disease model, and you have detected some change in this gene product level or activity with the disease, then you have a real, nice drug target.

Between these two views lies reality, and when companies get fed up by being told that they cannot go beyond a physiological target like insulin, they embrace the molecular biologists' view. Once they have been burned a few times with these "only molecular biology" targets, they start to paddle back. Meanwhile, a number of companies make it, and a number of companies go under.

Following Old Comfortable Formulas

If you ask senior people in Pharma companies what they would like to achieve, they might say they would like "to repeat success." Therefore, it is extremely important in every drug company to know the history of successful projects. For the most part their history is in their company's **chemical library**.

The chemical library of a company comprises the real and tangible history of all the projects they have worked on, and all the compounds they made to hit those projects' targets. So, if you work at Hoffmann La Roche which invented Valium and many other benzodiazepine drugs, then

[95] That is, how many receptors, receptor subtypes, their anatomical distribution, and how many of these insulin receptors are hit to produce the desired therapeutic effect when we add insulin with simple replacement therapy?

a much larger percentage of the compounds in the library are benzodiazepines than, for example, at Abbott. As a researcher working on the central nervous system (CNS), you would almost always have "hits" in a Roche-type library as Roche has worked on CNS targets for a very long time. If you worked in antivirals, you would have many fewer hits because Roche has a shorter history of working in antivirals even though it was one of the earliest HIV drug companies. From a business point of view, perhaps one should be aware of, or rather beware of, companies entering a new field unless they acquire a larger and more diverse chemical library to back up the strategic initiative. Mergers' first real result is the immediate merger of chemical libraries—just before the merging of sales forces—long before the merger of procedural standards of evaluating so many questions about clinical candidates and development plans, and much, much before the merger of company strategies and cultures that may not even happen before the next merger takes place!

Companies also have a psychological history. The company's personnel who have actually made a drug for the company and have lived the experience of successful projects are held in very high regard and command the greatest respect of their colleagues in the company, most of whom have never successfully discovered and launched their own drugs despite equal talent and effort and equal number of years spent at the company.

The other thing executives with an eye on avoiding risk are persuaded by is the apparent greater success with the drugs that are "followers" or "me toos" aimed at a validated target. Somebody—in extreme examples the company itself—has a clinically efficacious molecule, and the company tries to make a little bit better one. Of course, this is a "self-fulfilling prophecy" if all your efforts are only on noninnovative projects. The industry has only a few, often frustrated, "pioneers."

When it comes to target types, there is an even stronger collective memory. The collective data from the industry collected and analyzed by themselves and by the two to three major consulting firms (Accenture[96] and McKinsey) indicate that success comes from the following targets in decreasing order: G-protein coupled receptors (GPCRs), enzymes, ion channels, nuclear receptors, and so on (see Figure 12.1 for details).

If only for the reason that the industry will run out of functionally appropriate GPCRs, there will have to be increasing success with currently little exploited categories of targets including protein–protein

[96] Formerly, at the time of the survey, Andersen Consulting.

Target Types

- GPCR (small ligand)
- Enzyme (small substrate)
- Ion channel
- Nuclear receptor
- Protease
- Enzyme (large substrate)
- GPCR (large ligand)
- Cytotoxic/other
- Protein kinase
- Protein–protein

The industry has often continued focus on target types that have brought rich rewards in the past.

Figure 12.1 Successful Approaches to Drug Discovery in the Pre- and Postgenomics Eras

interactions. In 2003, "nutlin" was described by Roche as a very successful example of targeting protein–protein interaction. All of a sudden, medicinal chemists no longer think this is not doable. Driving this will be new discoveries from the genomics era that will assist the discovery of new targets whether receptors or other molecules.

For the immediate future—at the time of writing—if a scientist working with his or her microarrays from tissues in an inflammatory state comes up with legitimate targets comprising, say, 1 GPCR, 15 protein kinases, and 20 nuclear receptors, one can be sure that the company will pick the GPCR to work on. But it is going to be very difficult to engineer selectivity and find a clinical candidate that doesn't cause side effects by acting on similar GPCRs. Yet it is likely that there will be hits when screening against the chemical library, and thus medicinal chemists have a starting point. In the absence of hits, drug development projects cannot get started, or just those very hard ones using design-based *in silico* modeling on biological structures.

But it is hard to argue against a modest upside being strategically better than a precipitous downside in this risky and very capital-intensive business.

Targets from Clinically Broadly Active—or Dirty—Drugs

The concept here is to try and make better drugs from a very effective drug with multiple effects. You do it by dissecting the multiple actions of clinically active drugs. Unfortunately, it is not easy to describe this without continued reference to rather technical concepts.

For example, benzodiazepines are robust, efficacious anti-anxiety drugs—anxiolytics—but they are a sedative, interact with alcohol, cause dependence, and show tolerance development. They weren't discovered in the era of molecular biology, and, so, it has become known only more recently, long after they were discovered, that they act by being so-called allosteric ligands to an inhibitory ion channel receptor, the heteropentameric $GABA_A$ receptor. It comprises five subunits that form a chloride ion channel that opens upon binding of the common brain transmitter gamma-amino-butyric acid (GABA), causing a decrease in excitability. It has a separate binding site for these nonnatural molecules, the benzodiazepine-type drugs. The benzodiazepines simply increase the affinity for GABA, making it work more effectively and the cells less excitable. It is not a surprising mechanism or mode of action for a tranquilizer.

To make it more complex and simultaneously interesting, the $GABA_A$ receptor's five subunits can come from any one of six α-subunits and β-, γ-, and δ-subunits. There are a lot of potential structures for the receptor, and these are difficult to separate.[97] This presents a challenge as one tries to elucidate which particular receptor configuration lends the benzodiazepines their different effects and side effects. Only in the last five years or so have high-throughput screening (HTS) assays been used to pair the structure with the effect at Merck and a number of other companies. It is quite clear that the receptor subtypes are now being sorted out—by cloning and expressing different subunit combinations. Of the naturally occurring receptor types that emerge, some of them will carry the sedating effects of "benzos," and some of them will carry the anxiolytic effects. Hopefully, these will not be the same, and thus a subtype-specific drug may be as robust an anxiolytic as Valium without its sedating and other side effects. Indeed, several companies target $GABA_{A\text{-}\alpha 2}$ subtypes now for a pure nonsedating anxiolytic.

In fact, new effects, earlier masked by the simultaneous action of the nonsubtype-specific drug can also be discovered this way and $GABA_{\alpha-5}$ receptor antagonists are now in trials for cognitive improvement. Thus this approach may uncover new targets.

---

[97] Making an assay for the different heteropentamers is a "nightmare" because scientists are not terribly good at obtaining one given stoichiometry of the various possible pentameric ion channels. Working with the α-1 or α-5 subunits is fine, but if you want a combination of α, β, γ, and δ subunits and you want to mimic the different subunit combinations which you find at different sites in the brain, and which mediate different effects, it is pretty tough.

Box 12.2 Pathways to Drug Discovery: β-Blockers, ACE Inhibitors, and AT1 Receptor Antagonists. Three Classes of Antihypertensives

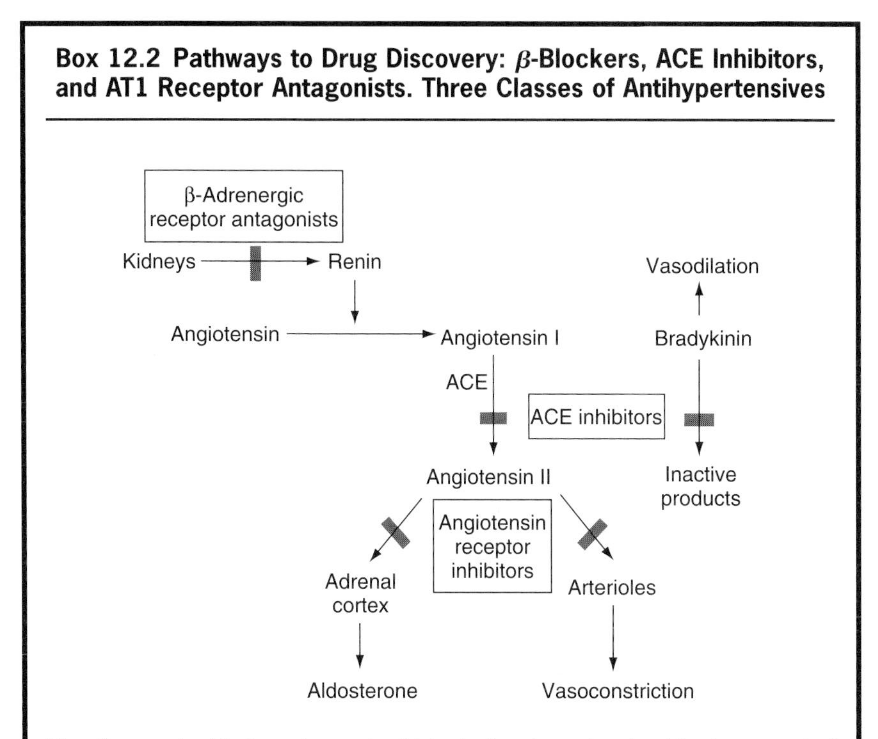

The diagram in this box shows the biological pathway involved in the control of blood pressure. In terms of drug discovery, the ***pathway*** is validated by the response to various drugs. Angiotensin receptor (AT1R) antagonists were clinically validated by the success of ACE inhibitors that showed the blood pressure lowering is achieved if the angiotensin II concentration is reduced, and thus fewer angiotensin II receptors can be activated. Both drugs reduce the concentration of the biologically active angiotensin II (AT1R-ATII complexes), and thus AT1R antagonists will reduce vasoconstriction and blood pressure. There is no practical distinction between the drugs that are enzyme (ACE) inhibitors and the ones that are receptor (AT1R) antagonists because basically the drug companies have made something that acts on the angiotensin pathway. However, the ACE inhibitors and AT1R antagonist are very different chemical structures and could be independently patented. Since the metabolism and some of the side effects may be different, some doctors and patients may prefer one to the other while exploiting the same mechanism that through AT1R lowers blood pressure. Many patients with high blood pressure are on several drugs because the maximum effect from a given biological mechanism is limited. Since blood pressure is regulated by several mechanisms simultaneously, one must be aware of the potential issue of drug interactions in the design of antihypertensive drugs. Many spectacular recent failures emanated from failing to address this issue.

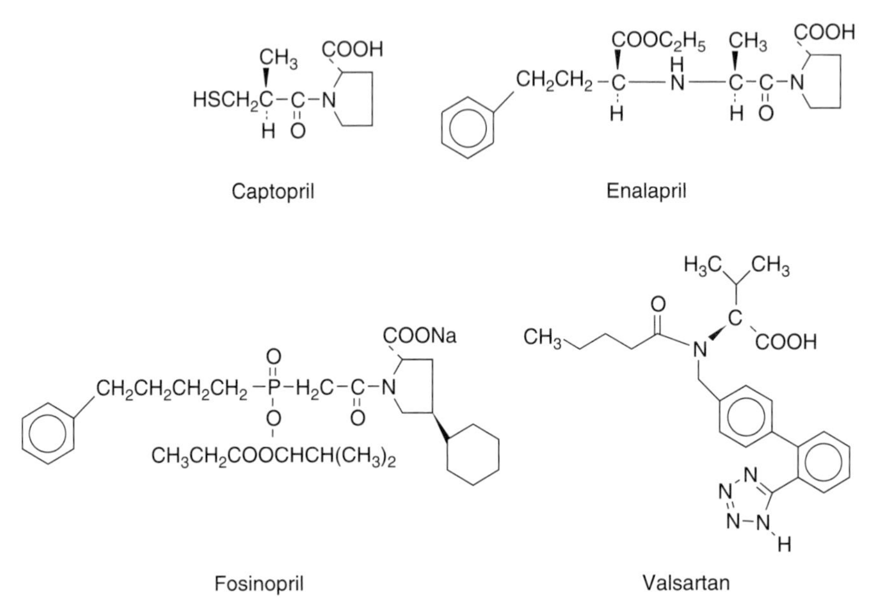

These drugs are all angiotensin inhibitors and all successful antihypertensives. Captopril, enalapril, and fosinopril are ACE inhibitors, and valsartan is an AT-1-R antagonist.

Figure 12.2 Structures of some ACE Inhibitors, and AT-1 Receptor Antagonists

chapter

13

PRACTICALITIES: THE HOOPS AND HURDLES OF BIG PHARMA

WHY IS THE SELECTION OF A CLINICAL CANDIDATE SUCH A PROBLEM?

Why is the selection of a clinical candidate such a big problem? At the start of a project people sit down and say, "*Wouldn't we love to cure Alzheimer disease?*" And everybody says: "*Yes! That would be great! Let's cure it! It is such an ugly neurodegenerative disease, and it is big and growing.*" Then the company will agree on what the drug should preferably do. It should: slow disease progression; or, better, it should stop the disease progression; even better it should start to make neurons grow back; and it should do it preferably in a short time, safely. This is very easy to agree on. But all too often, by the time the compound is made, it does none of the above or does not do all of it.

No clinical candidate molecule ever fulfills all of the drug–target profile requirements in terms of efficacy, safety, or ease of administration originally agreed on by the scientists, clinicians, and marketers. The company is faced with the reality that it has spent a huge amount of money; it would love to be in the Alzheimer disease market; and the company has to be in central nervous system (CNS) diseases if it wants to be a big company.[98] But the drug it has is not exactly the one it wanted. The company is in a dilemma. It does not have the drug that it knows would "sell itself." How much can the drug actually deviate from that which "the doctor ordered"? What is a company to do? What makes all decisions difficult? Reassuringly for the marketers, the competing companies have also not made the perfect drug either; if they had no one would believe them. And remember, both the clinical landscape and the competitive marketing landscape are changing during the years it takes from the initial agreement of what to treat and how to treat it to the selection of the clinical candidate. In addition, the people in marketing are likely to have changed twice during the development of the drug.

We have discussed many reasons to stop a drug discovery program. Clinical research is 5 to 20 times more expensive than preclinical research. The liabilities in clinical research and in selling medicine are very large. More importantly, the heads of clinical research have already budgeted for failure of a certain number of clinical candidates and INDs. But there are very persuasive reasons for pressing ahead even with a less than perfect candidate.

THE PIPELINE

Pragmatically, the companies are geared up for a steady flow of new products, which means a steady flow of clinical candidates to put into trial. Capacity at CROs helps to "even the flow." If a company is not putting drugs into clinical trials, confidence from investors will drop, and they will adjust their portfolios accordingly. Stock price is as heavily affected by the prospects of a company's pipeline as by its current sales.

It is very tempting to launch drugs. One starts with virtually valueless chemicals and manufactures a product that is worth on average 24 times its weight in gold. The Pharma industry is the epitome of added value. Sometimes the clinical candidate looks really good now that you have to make it in kilogram quantities to enter clinical trials. Many times the synthesis has to be reinvented for the larger scale manufacturing because it has dangerous steps, expensive starting material, or effectively no starting material at all for such quantities. If your R&D team has started out with a natural product where the annual world production is 8 kilograms and which has to be extracted from a bamboo that is reserved for pandas, then you have a problem with the manufacturing of your product. The solution

[98] Without cardiovascular and CNS programs, it is very difficult to be among the top 10 companies in terms of sales.

in that case is provided by the chemists who must come up with a different synthesis. Great natural product chemists—as consultants—can charge a lot for this service. Incidentally, it is important to charge a lot because of the maxim "the less you charge, the less they will listen."[99] Every consulting firm knows this, but few of the academic professors recognize it.

Marketing Machinery

Every time a company wants to announce that it will launch a drug, it doesn't talk about its research. It talks about how many more thousands of sales representatives it has hired to visit physicians. Once it has this "marketing machinery," it must use it. If a company has a broad program in, say, cardiovascular drugs, like Pfizer had, it could not afford not to own a cholesterol-lowering drug like Lipitor. Its sales force needs to sell everything the cardiologist needs. If a company cannot make a competitive, effective drug itself, it licenses one from someone with less marketing muscle. Or, if that does not work, the company with large marketing prowess buys the company that makes the drug it needs. Pharmacia-Upjohn-Searle (Pharmacia) let Pfizer sell their COX-2 inhibitor Celebrex because Pfizer could compete in marketing better against Merck's Vioxx. Pharmacia, then ended up being bought by Pfizer. Earlier, Pfizer lacked a cholesterol-lowering drug in the cardiovascular drug panel offered by its reps visiting the cardiologists and GPs, so it licensed in Lipitor, and later bought Warner-Lambert. The model was reapplied for Celebrex: co-market a drug and then acquire the owner.

Given all this investment, the company is receptive to new candidates from its R&D team. Assuming the R&D team has done its job and is proposing a candidate that makes sense, the company is receptive. The compound has to be potently efficacious and selective with few side effects, none of which are serious, and it has to be safe and stable on the shelves of hospitals and pharmacies. The clinician in the group brings to the planning meeting a biostatistician, who can be the scientist's worst bad dream, because the biostatistician will determine how many people you will have to enroll in the trial to prove efficacy. Biostatisticians have a tendency to overestimate this number: "40,000 patients should be enough!" A poor biostatistician, like the bad toxicologist, can be the kink in the armor of the company because, given half a chance, they will effectively price the potential drug out of the market by making the trial an insurmountable obstacle

[99] A new enterprise called Innocentive (https://www.innocentive.com/) has set up a system where especially problems of synthesizing pharmacological candidates is presented as a competition with prizes for the winning synthesis. There may be others.

at the planning stage. Nevertheless, underpowering a study by recruiting too few patients will prove costly in two ways. If you are unlucky and it's a long trial in the elderly, many simply may die for other reasons so that you will not have enough data at the time the endpoint is to be read—say 24 months after the start—and you will have to restart. The lost time and patent life—potentially two years each of $1 billion of irretrievable income—is substantial and your entry to market ordinal may be increased—first to third in class, say—so marketing will be a more costly uphill battle. Sometimes taking too small a group of patients backfires in an even worse manner. Assume, as happened to several big drugs at first, that there are numerically more emerging cancer cases in the drug treated than in the placebo group. The drug company is convinced that this finding has nothing to do with their drug's action; it's just chance because statistically the trial groups were small. Nonetheless, either they accept a label from FDA: "This drug may cause cancer" simply because more people had cancer on the drug than on placebo, or you redo the study on a large enough population in order that any chance appearances of cancer equal out. Now you have lost one to two years and paid for a second larger trial, as you should have done from the beginning. Knowing all this, one does not argue with the biostatistician.

Navigating through the Clinical Candidate Selection Board at Big Pharma

What is it like in the decision chamber of a Pharma company? Each of the therapeutic area heads has prepared for about six months. The scientists' bonuses are dependent on his or her drug candidate being developed.

What would they have decided about the real clinical candidates we have discussed in the prior chapters? The scenario is about the same for all companies whether in New Jersey, Switzerland, or the UK. The ground rules are very clear: "*You cannot send your deputy, you cannot bring an outside expert, no matter how many Nobel prizes he or she has, and all the documents have to circulate at least 14 days in advance.*" Last-minute additions to the agenda or pronouncements along the lines of: "*Yesterday, we got this 'phone call from clinician χ who has said that there was a 'miraculous effect' using our candidate*" are banned. This does not mean that not all of these can ever occur, it only means that they are not supposed to occur. One should be aware that considering the limited development budgets and the competition for them, *people* within the company are the worst enemies of any drug candidate. Even though, in a way, it is in everyone's interest that the most likely candidates should go to the clinic, the choice of clinical candidates can be very competitive within a company. However, providing the company still has profits from drugs, it is of no matter to the CEO or the stock market if the profits come from drugs possibly developed by people long retired, fired, or otherwise forgotten.[100]

Perhaps in reality, although none of the rules is negotiable, all of the elements of the drug developmental process may be compromised every time. There are no perfect molecules for clinical candidates. Moreover, the hope is that the next generation of drugs will be better—and they will be or will not be released.

What is the most important item of all (see Table 13.1)? It is: "If you don't know how to do the clinical trial of your clinical candidate, then you don't have a clinical candidate." This is really the most important point for all scientists to understand whether they are in biology or chemistry. There are two kinds of company. Unfortunately, in most companies people are so busy being chemists, or toxicologists, or biochemists, or assay developers, that they

Table 13.1 Clinical Candidate (CC) Selection Board of Imaginary Pharma, Inc. of Newtown, New Jersey, Baseva, Switzerland, or Essex, UK

Ground Rules

- No deputies, no outside experts
- All documents must be circulated 14 days in advance
- No last-minute additions to the agenda

Directives from the Board of Directors Meeting

- If you do not know how to do the clinical trial of your CC—that is, which patient group, how large, how long, how to administer the drug in which doses—you do not have a CC.
- If you do not have biomarkers and surrogate markers, and dosing suggestions, you do not have a CC.
- If you do not have the buy-in of the U.S. marketing organization, you do not have a CC.
- If you do not have 10 or more years of patent life remaining, you do not have a CC.
- If you do not have a backup compound, you do not have a CC.

Questions to Be Answered

Is this a significantly better therapeutic than those available?

Will we be able to show it to the satisfaction of the FDA, to have it approved, and send to the market to have it distributed and sold at a good price?

[100] Somebody compared this scenario to the concerns of a minister of defense. The country does not really care if you win the war with a strong Army or strong Marines as long as you win, but the Army chief and the Marine chief see things differently, in particular in peacetime.

never take the trouble to understand the clinical trial or in fact all the possible clinical trials their compound could be or will be tested in. In the other, regrettably rarer, kind of company everybody, even a chemist who is just given a structure to synthesize—"we need this structure as a pure stereoisomer not in a racemic[101] form and we need it by next month"—understands why. They know for which indication this compound is intended, how the trial will actually be done, how the competitors' drugs work, and what it will take to be better. If anyone in the team does not understand the bigger picture, they will not be helpful to anyone. After all, NASA's cleaning lady knew that she worked "to put a man on the moon," so NASA could do it.

Indirect Treatments and Surrogate Markers

As we have explained quite often, the developmental trick is that in many (but certainly not most cases) the indication the clinical trial is designed around, and for which the drug is approved, is not what the drug will be mostly sold for if everything goes well. Sometimes it is to test the waters for a compound with a new mechanism of action that an indication is chosen. A smaller and shorter (i.e., cheaper) trial can let the company convince itself that it is safe, it reaches the target, and it will be sensible and eventually profitable before taking the compound into a large, long (two to three years), and very expensive ($200 to 300 million) trial in a large indication. Making two to three mistakes of such trials in a year would in a few years wipe out even the largest companies, so the caution is understandable. There are many ways to show that your drug is efficacious and safe, and the key is that "**safe drugs will find a use**." But this is not always in a predicted, contrived, or even straightforward way. When Mr. Schwartz is stopped in the street by his physician, and the physician asks: "*Mr. Schwartz, how are you doing on the new drug I gave to you against urinary incontinence? Does it help? I am happy to see you on the street. You told me you did not dare to leave home.*" Mr. Schwartz turns around. He says: "*Your colleague put me on an anxiolytic, now I don't care.*" There are many ways to reach an effect. If you are involved in drug discovery whether as a consultant or by running your own company, you should insist that the team explain to you how they intend to test their compound in the clinic. If they do not have a clue, the company doesn't have a drug. It will never have it; there is no way! Bringing in people

[101] Synonymous with enantiomers and pertaining to *racemate*—an equimolar mixture of a pair of enantiomers—respectively. See for example http://www.chem.qmul.ac.uk/iupac/stereo/RS.html#47

to determine how to design and execute a trial when the compound is ready, usually leads to expensive and failing trials. Yet most Biotechs plan to contact clinicians and regulatory experts after the first three to four years in business, when they might have already burned $30 to 50 million.

The key elements of clinical trial design are the same. You have to have a very good estimate of the effective dose. The scientists know that the concentration of drug needed to inhibit the enzyme sufficiently in the tissue is, say, 10 micromolar. But what dose of drug needs to be given and by which route to reach this concentration? One cannot take a biopsy to measure the concentration. And if you are performing a trial to treat depression, you cannot wait months to find out what the dose *should* have been. If you do an obesity trial and you have to show that the initial weight loss is retained, you will do your trial for 12 to 18 months, but you need to know very early in the trial if you are using the right dose. Sometimes it is quite hard, for example, in the area of pain treatment. So far no trial has been successful with a fixed dose of an analgesic drug, but rather by supplying it within the limits of safety to be taken by patients to reach the desired pain relief without exceeding the maximum dose. The reputed gold standard gabapentin and all other major pain drugs were shown to be efficacious by this dosing method.

While the pain patient knows she is in pain and can adjust the dose, the epilepsy patient does not know that he is building up for seizures three days from now by underdosing, which may have been caused by a simple bout of diarrhea that removed the drug faster than planned. The depressed patient who is told that it will take 14 days for improvements to be noticeable cannot know if he is being underdosed in the first two weeks, or the lack of effect is simply the time it takes for the effect to build up. Once again, you need a "surrogate marker." There are many ways to do it, but this is an ingenious and graphic example. If you, for example, are making a new antidepressant which is one of the new—very new—*substance-P* antagonists, specifically an NK-1 antagonist, and you want to know what the concentration is in the brain, then you explore this in the indication *emesis*[102] (i.e., you induce vomiting and then try to block it with the same substance-P antagonist). A substance-P antagonist can indeed block vomiting through a central brain mechanism, so this gives a good indication of dose vs. effective concentration in the brain. So, in 10 unpleasant minutes, you discover how many of the volunteers still throw up because they have too low a concentration of the NK-1 antagonist in the brain. If you have the dose right, they do not throw up. It's a very graphically qualitative assay; you don't miss it.

---

[102] [Induction of] vomiting. Emesis is an unfortunately common side effect of chemotherapy of cancer—hence the great need for effective antiemetic drugs.

Perhaps another lesson from this example, and others like it, is that not enough companies actually work on these small indications like emesis[103] or, in Mr. Schwartz's case, on the growing indication of urinary incontinence. These are examples of extreme medical problems that nobody really works on seriously, just opportunistically. If a drug designed for a large indication such as depression also happens to work in another therapy area, companies can produce a separate formulation with alternate pricing for the opportunistic use. It is even better if you have a good backup compound that basically works like the tested one. Then the marketers will sell one for the small indication and the other for the bigger one without headache over how to price them. It's a huge opportunity for small companies.

Three Most Important Things: U.S. Market, U.S. Market, and U.S. Market

Today, 62% of the sales of Pharma is in the United States. But how many know that there are over 800 compounds that are sold in Europe, and which are highly efficacious, therapeutically wonderful, but which have not yet been registered in America? And they won't ever be. Because by now the patent life is so short, and the FDA is so slow, that they cannot be. Every company understands that unless the American marketing arm of a multinational company says: "It will be marketed in the States," there is no real point even to make the drug. If, for a variety of possible reasons, a company doesn't have a lot of patent life left, then the marketers in the United States won't be interested. And it is nonnegotiable; this is how it is. Sometimes new formulations and other modifications help a good European drug to enter the United States and make a real cut. Memantine has been used since late 1980s in Germany as a neuroactive substance in the treatment of Alzheimer disease patients. Now Forest Laboratories, a pioneer in picking up good European drugs, for example deprenyl (Selegiline) for Parkinson disease, and cipramil (Celexa) another SSRI for depression, will start to sell memantine[104] as Namenda in the United States for treatment of moderate to severe AD. It was worth rerunning

[103] However, in this case, NK-1 antagonists are in trials for emesis as well as major depressive disorder (MDD) and, incidentally, pain, with which substance-P has been associated since its discovery.
[104] From http://www.frx.com/careers/prodmarkets.shtml (contd.)

trials once they knew that a certain dose worked in German patients. It became the first in its class of drugs to slow disease progression in AD.

THE IDEAL BACKUP COMPOUND

The other directive, "If you don't have a backup compound, you don't have a clinical candidate," is very important, but drug companies often make the mistake of going ahead without a backup in the works. What is the ideal backup compound? Ideally, it is acting on the same biological target, with the same efficacy and selectivity, but is of a different chemical structure class. This in itself would help scientists to decide between mechanism and molecular structure-based toxicity (i.e., are the side effects linked to the target, or are they a result of the chemical structure with which we approach this target?) If the latter is the case, then it is worth going back to the screening hits and developing another chemical class of compounds to reach the same target. However, in reality many times the backup is of the same structural class either because there were not many hits in screening, or because it was easier on medicinal chemistry resources. The backup should be ready in time for the trials so that it can be switched if necessary for the first compound, and those patients recruited for Phase II and Phase III are not missed.

If there is more than one indication for a ligand the different structural class becomes particularly attractive. Incidentally, if both the original and the backup reach the market having two different structures for two different indications, it will also help in marketing especially since the company will not have a problem with pricing it. From the current example, NK-1 antagonists are in trials for major depressive disorder (MDD), pain, and emesis, but they are all of the same class of compound.

Even small companies should have a backup, but they more often do not. When they lose their only compound in Phase I, they have often nothing left, whereas any medium to large Pharma has 20 to 30 ongoing Phase I trials.

---

[104] contd. Forest is currently developing **memantine** for the treatment of moderate-to-severe Alzheimer disease. Forest originally filed a new drug application (NDA) with the FDA in July 2002 but withdrew the application because new data from a late-stage study reveal that memantine has more potential than is found in initial trials. The results show that when the drug is combined with the Alzheimer's treatment donepezil, it offers major benefits to patients when compared to the use of donepezil alone. Forest amended its application by the end of 2002. This application included the study's efficacy and safety data for memantine as a single therapy plus new data for a combined therapy. Forest is conducting additional trials further to confirm memantine's cognitive and functional benefits and to evaluate its efficacy in patients with mild-to-moderate as well as moderate-to-severe stages of Alzheimer's. Memantine is currently marketed in Europe for the treatment of moderately severe-to-severe Alzheimer disease.

Why Are Clinical Candidates Being Selected at All?

If it is so difficult to select a clinical candidate, how are they being selected at all? Primarily because the medical need is there. Secondarily, because it is a necessary step in making a profit and staying in the business. Yet moreover and most sincerely, because the preclinical scientists and the clinicians truly want to make drugs that alleviate suffering and, hopefully, this is not going to change. They are trained to determine if there is a new therapeutic possibility. Which candidates and indications are chosen changes continuously as the market changes sometimes through the lobbying of patient groups and advocacy of Congress and other bodies.

Ultimately, with a stringent set of animal toxicology data, the FDA will grant the companies' clinical candidate an investigational new drug (IND) status that allows it to be "entered into man," that is, tested for safety in a Phase I trial to be followed by efficacy trials.

There are no perfect molecules, and for the company the ultimate question is: "**Is the molecule you have better than those available, and can you show this to the market and the FDA?**" This is truly the matter.

chapter

14

PRACTICAL TRIALS FOR A BALANCED PORTFOLIO

IS YOUR CLINICAL CANDIDATE SIGNIFICANTLY BETTER?

The question is: is the molecule you have as good as or better than those available, and can you show this to the FDA? The FDA is looking for other ways to evaluate drugs by examining different endpoints for trials, for surrogate and diagnostic markers, which will have the consequence of stratifying patients to such an extent that the patients enrolled in the trial are core representatives of the disorder. The design of trials to satisfy the FDA and other bodies is a constantly moving target. New tests are added—such as a bone density scan instead of waiting to count the number of hip fractures in osteoporosis; new neuropsychological battery tests in Alzheimer disease (AD)—that on the one hand make it very difficult for companies to develop a standard clinical trial routine for a disease. On the other hand, many of the new markers, despite adding costs, are revitalizing drug discovery because they provide surrogate endpoints that help the experts and the company to judge earlier if the trial is going well.

No patient pays for increased bone density. However, they would pay to have fewer hip fractures, but the correlation is so good that the company whose drug stops the decline of bone density at six months knows that it is most likely to have fewer hip fractures by the end of the two-year trial of

1,000 postmenopausal women—an often selected patient group in osteoporosis trials. Nevertheless, for approval the FDA still wants the two-year hip fracture data too, until it has been shown for a dozen drugs that bone density maintenance **always** reduces hip fractures. Although insurance companies and health administrators may have been upset at the expense of the bone density measurements, it increased the number of drug candidates in osteoporosis. This is a good result for society. In the long run, when drugs are cheaper, more patients will be helped without having to undergo hip replacement. Perhaps surgeons and artificial hip manufacturers would still lobby against Fosamax and other drugs to treat osteoporosis.

Other new surrogate markers and new diagnostic tools make drug development better but almost always costlier as well. Take the example of AD. Today signs of dementia can be diagnosed very early with 95% accuracy by good standardized neuropsychological tests. So if you are to test a drug that slows disease progression, and you start with these early AD patients, their decline is slow and, therefore, even a very efficacious drug will take 24 months of treatment to show a difference between placebo and drug treatment. This is not enough for the FDA. To be rigorous, you might then be obliged to stop giving the drug and give placebo to both groups to show that the rate of decline now is the same as in the placebo-treated group, and then reinstate the drug to see that you again can slow the disease.

Many things in clinical trials could be improved. After a hundred trials on antihypertensives, trials still have placebo groups, and companies cannot reference their drug against historical data from a "statistical placebo group." This is despite the fact that the industry probably has measured and reported to the FDA some 10 million blood pressures in placebo groups. Using historical data in this way would not only be a cost saving to the company, since the prime advantage would be that all people in the trial would be on the drug being tested. Society loses when the hurdles are set too high, just because they can be.

In addition, the length of trials is different (antibiotic clinical trials can last 14 days, but a new Alzheimer drug trial would be 24 months, and osteoporosis drug trials are 24 to 48 months). And the size of trials is different. Is there a sufficient number of patients who are well characterized and who can enter the trial so that there will be the statistical power to show efficacy and it will not take forever? In cardiovascular medicine it may require 30 to 400 patients if it is a drug for heart failure, 1,000 if it is for high blood pressure, and 3,000 if it is a cholesterol-lowering drug. In stroke or sepsis, you might need a minimum of 60 to 100 patients, depending on the degree of improvement one wishes to demonstrate and on the natural rate of disease progression. In obesity you may need 2,000 to 10,000 patients. If that doesn't already present sufficient problems, the inherent risk to both companies and patients of trials is different. At worst, the mortality of the patients can be significant, and this is always a matter of great concern. In a given Big Pharma company maybe 30 to 40 Phase I-III clinical trials are ongoing, and the "ups and downs" get diluted. In Biotech, the ups and downs are more pronounced as most of them have one to two drugs in the clinic, and we know that on average of those that entered clinical trials only one of ten drugs will make it.

Since clinical expertise is the key in successful trials, if a company lacks experience, joining forces with another company may be the key in successful trials. In this way the much smaller Eisai used Pfizer for Aricept, presently the most successful symptom treatment drug of Alzheimer disease.[105] Small companies can share the cost and risk of clinical trials with big companies, and, of course, tolerate the loss of most of the upside. The critical notion is to make sure that if you are looking for a partner, you as a company should license the product of your hard preclinical research not to the highest bidder (as Biotech tends to do) but to the most experienced company in the field. Sometimes the experience may be in understanding the regulatory system that can be key to rapid approval with "good label." In this way Astra worked with Merck for omeprazole, one of the largest selling drugs ever once Merck had helped its registration, introduction, and marketing in the United States.[106] Even the task of manufacturing your candidate drug in sufficient quantities for large human trials is demanding and is better shared.

[105] Aricept or donepezil is used in the treatment of mild to moderate dementia in Alzheimer's patients.

[106] Used to prevent ulcers and to treat other conditions where the stomach produces too much acid.

chapter

15

How to Improve the Odds of Finding a Safe Drug That Works

Therapeutic Windows

A functional definition of the drug target is: the specific site at which the therapeutic actions of a medicine are exerted. All the other sites at which the drug may act are, for all discussion and approval purposes, "disreputable" because they are, or at least may be, the sites where the side effects arise. If there is a *big* difference in the affinity for the target site and the side effect site, in favor of the target site, then there is *a big* "therapeutic window." With such a large therapeutic window, one can increase the dose, 5, 10, even 50-fold, before having serious side effects. Ideally, you as drug developer, physician, or patient would like to know that the drug that the patient takes has a 100-fold or 500-fold therapeutic window between hitting the real target and other targets. Remember that the dose for a larger group has to be determined, already taking into account that interindividual variation in drug metabolism can be as great as 10-fold. So a therapeutic window of 10 is no window at all for some patients.

However, about 90% of all old drugs have a window that is less than 10 to 20. This means that if somebody goes home and starts to take an overdose, then he or she very rapidly will incur and endure possibly serious side effects.

The undeniable truth is that in complex diseases, where we do not fully understand the disease mechanism—depression, anxiety, schizophrenia, rheumatoid arthritis—it is terribly difficult to separate all positive, therapeutic effects and side effects. But it is partly of the drug companies' own making. It is a direct result of drugs being developed for very highly distributed and active targets of particular classes of proteins such as the G-protein coupled receptor (GPCR), and ion channels abundant in the brain, spinal cord, and heart, and the proteolytic enzymes such as the caspases. If you make drugs to the same class of targets all the time, on the one hand, they are a known, familiar entity to the companies' biologists and chemists and one knows what one is doing. On the other hand, it is very hard to engineer selectivity. Moreover, the classes of drugs chemists use to target targets are, as we have emphasized, not very diverse—as we discover when they run into each other's patents quite disturbingly often.

Statistics of Failure

Major Pharma doesn't like to distribute its data on failure to stock market analysts, lest they pull down their ratings, but they can give them to a common—in the sense of shared—consultant. The combined data from the top companies (see Table 15.1) show that failure is industrywide; no one player in the industry is significantly better than the other. But different indications have different failure rates for different reasons, but failure rates are not different between the companies for the same types of indications.

Antibiotics virtually never fail on efficacy because the clinical candidate either kills the bacteria or it does not and this can be ascertained with 100% accuracy *in vitro* looking at bacterial culture at the laboratory bench. They fail on safety, or because the dose needed is causing too many unwanted side effects. Who does not have a really upset stomach on a broad spectrum antibiotic? At the other end of the spectrum of indications, CNS drugs fail mostly on efficacy during Phases II and III after costly large trials. Finding the right dose for CNS drugs has been greatly enhanced by PET imaging using the proposed drug. Now one can study where it binds in the brain when administered at the proposed dose, and how large a portion of the receptors are occupied. For the most used

Table 15.1 Some Statistics on Failure after Entry into Man

	Antibiotic %	*CNS %*	*CV %*	*Resp %*
Failure due to safety (Phase I)	85–100	20–30	30–40	50
Failure due to level of efficacy (Phase II/III)	5–15	70	60	50

Statistics of failure assembled from the historical data from the top pharmaceutical companies. Note that failure is industrywide, and failure is also indicationwide. Different indications have different Phase I, Phase II, and Phase III failure. Antibiotics fail in Phase I, and 90% of obesity drugs (not shown) fail in Phase II. Reasons for the failures vary between indications, but not between companies, since the quality of research and the stringency of selection criteria for entry into clinical trials between the top 20 companies are not significantly different. Small Pharma and Biotechs can more often select a candidate that is showing only limited efficacy in their models, or works in some but not in others. They have to take this risk because they have too few development candidates and the investors are sitting on the doorstep. Luckily, the safety requirements for entering into man are the same for small and big companies, for industry and academia, for nonprofit academic pharmacological research in man, and for drug development for profit. The FDA's major function is to regulate the clinical part of the drug development process while the agency is legally prohibited from regulating the practice of medicine. Key: CNS = central nervous system; CV = cardiovascular; and Resp = drugs for respiratory disorders, mostly asthma and chronic obstructive pulmonary disease (COPD).

SSRIs, the number of trials was seven or higher before there were three pivotal trials in which they were shown to be better than placebo. It is good to be aware of this in advance before entering into any antidepressant development, lest you be discouraged too early or, worse, you allocate too few funds to demonstrate efficacy, or, even worse, you have underprepared your colleagues within the company. Some of your colleagues may delight in killing your project in order to have sufficient funding for their own development candidate. Educating your own company about the clinical trial risks is the most important thing next to understanding what the FDA will consider as a successful trial.[107]

The data for failure rates of cardiovascular (CV) drugs are similar to those for CNS, whereas for drugs against respiratory disease, such as asthma and chronic obstructive pulmonary disease (COPD), failure is shared 50:50 between safety and efficacy.

It is not unreasonable to extend this analysis and conclude that the quality of research is not significantly different within the top 20 companies. An important observation from this is that Biotechs, if you look at

[107] Genentech made headlines in 2004 for a reported 3–6 month prolongation of life in an oncology trial. It gained approval from the FDA.

their portfolios, are projecting to have expectations of about 20 to 200 times fewer failures than Big Pharma. Frankly, it is hard to believe that the people in Biotech will be 200 times smarter. Their principals and heads of R&D are all trained in the same academic institutions, by the same mentors. They may, of course, be twice as smart but not 200 times! To underscore the point, ***if they are not 200 times more successful, their portfolios will not produce sufficient products for the companies to remain viable in the long term***. Sucessful biotechs immediately expand their portfolio of projects to meet the industry-wide risk of failure.

Ways of Playing It Safe

These data of high failure rates are in a context of already trying to play it safe. Executive policy is too often to engage in "me-too" and "follower" programs with minimal modification of others' molecules—often with judicious "patent busting"—where the target is clinically validated, clinical trials are agreed with the FDA, and the risks are small. The strategy is to go for a validated target with a new chemical class, and hopefully with a better, or at least different, side effect profile. If you can do this fast and enter the market as fifth in class, with strong marketing muscle your drug can make it into the top three.

Patenting strategies of different companies are overtly aimed to keep out competition. Big Pharma will often—too often—refuse to grant licenses to competition even for clinical candidates they are NO LONGER pursuing. Biotech, more than Pharma, patents, or tries to patent, the targets themselves. All companies would like to have an exclusive Intellectual Property (IP) patent on a chemical class, and on a given new chemical entity (NCE), but under current regulations, if you

Box 15.1 Me-Too and Follower Programs

- Patenting strategies of different companies are aimed to keep out competition:
  - From a target (Biotech more than Pharma)
  - From a chemical class (patent busting is an art form)
  - If you get an efficacious, original drug using a patented target they must give you a license but you will pay some royalty
- But even very closed related compounds can have widely different pharmacokinetics and drug interactions
  - All statins are so far safe apart from Baycol (with some 31 U.S. deaths)

have shown a new use for the original drug, the original company, after some pressure from the FDA, must give you a license to allow you to attempt to alleviate human suffering.

Coming up with new drugs within an already known class of chemicals—or filling the chemical holes in the original patent—is very important since even very closely related compounds can have widely different pharmacokinetics and drug interactions. Mostly, this makes for a better market penetration of the second or third drugs of the same class. The harsh edges are not that sharp; some side effects are gone.

Sometimes, however, being in the same class does not prevent a copycat drug form from getting into new dangerous side effects. For example, Baycol was linked with 31 U.S. deaths[108] but is the only statin—a class of cholesterol-lowering drugs established already for 15 years in the market in tens of millions of patients—with any reported safety problem.

Another tactic, though rarer because it is chemically difficult, is to combine two known effects into one molecule. Thus, venlataxine was developed to block both norepinephrine and serotonin uptake and try to beat the SSRIs at their own antidepressing game. An additional compelling reason for pursuing this avenue was the robust efficacy of the 40-plus-year-old tricyclics that inhibit both norepinephrine and serotonin uptake, but that had many unpleasant, but not dangerous, side effects like dryness of mouth. A less glamorous but very effective approach is to focus on the side effect profile of a known efficacious drug and reduce the most common cause of dropout.

The deceptively modern-sounding approach to playing it safe is to turn to "biologicals." Having identified a naturally occurring hormone, cytokine, or chemokine with therapeutic potential, you produce it by recombinant means. You will have a safe, efficacious, yet admittedly difficult to administer drug. It is going to dominate the market until someone comes up with an orally available, small molecule to achieve the same therapeutic effect. One is entitled to call it "deceptively modern" since in this category of hormone replacement therapy, insulin has been dominant in its market for over 80 years and we are still looking for a low molecular weight, orally available "insulin." Years of research—including research from some of the finest Pharma companies—have not replaced insulin as its own hormone replacement therapy. Surely we have come up with insulin sensitizers or insulin releasers, and other ways to deal with declining control of insulin levels. These are important drugs in the treatment and prevention of diabetes together with diet and exercise. What we have not done is to make a low-molecular-weight insulin receptor agonist to

---

[108] As of 2003.

replace injectable insulin with a pill. We have made a subcutaneous injection form, and we have made a slow-release formulation—"insulin retard"—and have developed other delivery mechanisms (transdermal patches and the like), but it is still the same protein.

The first recombinant drug, growth hormone (GH), was developed from technology from the U.S. company Genentech. Kabi, a Swedish company, through its very farsighted scientific director Bertil Aberg, licensed Genentech's technology to make GH in 1978. Not only did Kabi make GH, but it also effectively "made" Genentech by enlisting the small American firm to clone and express human growth hormone. The road to Genentech—stationed among the meat-packing halls and warehouses in poor South San Francisco—was quite literally paved on the occasion of the king of Sweden's visit in 1982. As a safe and efficacious drug, GH has found many uses. While it was approved for treatment of dwarfism,[109] it now has four times more sales for its ability to help patients recover from hip fractures. In 2003 Lilly (the oldest American insulin producer) controversially gained approval to market it for youths with "slight" growth retardation or "normally short" according to standard growth curves. The company "promised" not to "push" GH in this market of families who believe that taller children will do better in life. It is a curious decision in the climate of 2004 that this type of ethical control was left to the judgment of the company. It will be interesting to see how much of a lifestyle drug GH becomes. Parents should beware: GH's effects are irreversible; "shrinking the kids" only works in the movies. Since GH is being unscrupulously peddled on the Internet to rectify all sorts of self-perceived shortcomings, parents, teachers, and social workers should beware of kids' self-procurement and administration of this as much as other overtly illegal drugs.

GH behaves almost as a generic drug in the commercial sense since it is one of the first Pharma products for which companies, in this case Genentech, Lilly, Pfizer, and Novo, compete on price. This phenomenon has now been seen in marketing for both α- and β- interferon by Roche vs. Schering and Serono-Aventis vs. IDEC-Biogen, respectively. While the companies claim their products to be better—for example "cause the formation of fewer autoantibodies," which, incidentally, reduces the effective dose and ultimately may also have its own drawbacks—the companies are obliged to consider making them cheaper too.

The other broad approach, which is particularly attractive to Biotech, is to identify tumor-specific antigens and make human or humanized monoclonal antibodies to them to inhibit and reverse tumor development. The

[109] As caused by a deficiency in growth hormone.

additional benefits are that the oncology trials required are small, and Phase I is often jumped over or lumped as a Phase I-IIa trial for biologicals making development times shorter. Safety records are predominantly good, making these a darling of Biotechs. Unfortunately for the smaller companies, the Big Pharma's intent to play it safe has motivated and driven their broad entry into the biologicals market, and half of the new approvals in 2002 and 2003 for Big Pharma have been biologicals. The biologicals they once frowned upon and criticized, they now embrace. Big Pharma now even invests in companies that only make biologicals, such as the major stake of Roche in Genentech or of Novartis in Chivon.

But, as a word of caution, not all of these intellectually tempting "monoclonals with potential" will work, nor do all biologicals work out, or have an easy, or short, trial, or prove to be safe.

Some biologicals are used because of their enzyme activity: to be able to disperse blood clots, such as tPA and urokinase. Tissue plasminogen activator (tPA) does wonders in clearing the blood clots in heart attacks; it helps tremendously by rescuing much heart tissue, as does the later introduced and cheaper urokinase. tPA, where it works, works very well, and its beneficial effects in ischemic stroke are remarkable. But tPA is only useful in 2–8% of strokes. tPA itself can cause huge hemorrhaging; the patient looks literally and quite dramatically blue. It is not widely used in all hospitals in emergency stroke treatment, especially those outside the United States, simply because this drug-induced hemorrhaging is potentially lethal in the wrong patient. Where tPA is used, its success comes from those institutes being able rapidly—within three hours of the stroke—and reliably to differentiate between different strokes and predict accurately the cases where it will be effective. tPA needs very good neurosurgical emergency care physicians behind its deployment.

One can never be sure, and caution is required especially in treating patients who stand a good chance of long-term recovery without intervention, which is many times the case in mild stroke patients.

This is another cautionary tale from the inflammation story mentioned earlier. The decision to use interleukin-1 (IL-1) soluble receptor-II to control inflammation—which is a replica of the thinking behind the soluble TNF receptor Enbrel in that it binds IL-1, rendering it temporarily inactive—quite surprisingly increased mortality in a sepsis trial designed to be a short (but not fatal) Proof of Principle. No one knew that it would only bind *transiently* to the IL-1 and, later, would then release it. It effectively acted not as a neutralizer of IL-1, but an additional store or sink that was subsequently released with devastating effect. Instead of decreasing the amount of IL-1, it in effect increased it. Such a result is both disturbingly unpredictable and very surprising;

hindsight is the only indicator. Nevertheless, another go at the IL-1 system to produce a strong anti-inflammatory was successful. The IL-1 receptor antagonist—Kinneret—is now approved for treatment of rheumatoid arthritis (RA). Amgen[110] has both TNF (Enbrel) and IL-1 (Kinneret) drugs, and since it is known that TNF and IL-1 enhance each other's inflammatory effects, this combination of drugs may be very effective in treatment of RA.

Nerve growth factor (NGF), which enhances the regeneration of nerves in every animal experiment ever attempted, seemed like an excellent idea to treat neurodegenerative disorders such as diabetic neuropathy or amyotrophic lateral sclerosis (ALS or Lou Gehrig's disease). Although patients fought to be enrolled in the trial in the first place, none wanted to come back for their second injection. Patients dropped out of the trials because of the intense pain. During the Phase I trial the essential, and prerequisite for proof, double-blind protocol was failing because the patients knew from other patients to expect local pain if they were given the NGF. The trial could only be rescued by using a placebo of salt solution to create the same pain in an attempt to keep the study "blind." Why was this so painful? Because of something that was not known at the time: the target was not specific. The target receptor—trkA ("Track A"), which was thought to exist only on nerve trunks that were damaged and trying to regenerate—is also present on macrophages that can be activated by NGF. The activation included the generation of hydrogen peroxide, which caused burning pain.

The use of vaccines against all manner of diseases beyond infections from Alzheimer disease (AD) to cocaine addiction has been touted by many, and with good potential reason. The principle of vaccination has now been proven for 200 years. The rationale is that by injection of a properly chosen antigen, we induce the body to make its own antibodies "in house," and these circulating antibodies attach and make inactive encountered viruses or, more recently in this case of ***therapeutic vaccines***, any toxic protein associated with the disease or essential to the process. In the case of AD, an excess of the peptide β-amyloid-140/42 is essential for the disease progression. The β-amyloid vaccines have to progress a long way to reach approval. Some tragic progress has been made since the first vaccine by Elan caused neuroinflammation and killed several patients. When studying the brain of these victims of the trial, it was shown that the antibodies do enter the brain and even contribute to the decline of already existing amyloid plaques. So now many companies are encouraged to repeat these effects of a β-amyloid vaccine but, of

[110] After Amgen bought Immunex to get Enbrel.

course, without the increased mortality. The Proof of Principle for antibody-mediated removal of free and plaque-bound β-amyloid is in the autopsy data from the trial victims. Some people who were also in the AD vaccine trial would have died from natural causes unrelated to the trial, and the autopsies of their brains give equally or even more important clues to understanding the potential benefits and risks of AD vaccines.

In the case of cocaine use, various "evoking molecules" are necessary for cocaine to have an effect. Hypothetical vaccines in development might be antibodies to these, evoking molecules for cocaine's effects. If cocaine has no effect, it will no longer be "enjoyed" and will not be "pursued" by habitual users. Of course, this strategy is unlikely to be trouble-free. It is sobering and instructive to reflect that most of our presently used vaccines, especially that for whooping cough, would never be approved in today's litigious climate since they have side effects in 1 in c.4,000 of the population, which is a very high number when every infant is injected several times with a cocktail of vaccines. In contrast, no normal drugs are used by more than 5% of the population. Moreover, prescription drugs should only be used by those who need them, with a recognized benefit and quantifiable risk; vaccines are administered to those who might need them and where benefit is at best insurance against a future possible and nonquantifiable risk.

The great problem with therapeutic vaccines is that it is difficult to predict the antigen concentrations needed to reach a "meaningful titer" that is, a meaningful antibody concentration. It is also difficult to predict the consequences of immunization. The immune system has learned to limit any immune reaction to "non-self" type molecules in microbes or man-made vaccines, and there are antibodies—"secondary" or "anti-idiotypic"—to the primary antibodies evoked by the microbe or vaccine, respectively. There is risk for a generalized inflammatory response to any microbe or vaccine, such as is seen in "complement activation." The first Alzheimer disease vaccine trial was stopped on March 28, 2002 after three people died. The irony of this drug development is that the autopsy of these trial victims confirmed that the vaccine was working—that is, there were the desired antibodies formed, they indeed reached the sites in the brain where the toxic β-amyloid was being formed, and the antibodies were present on the amyloid plaques that are so characteristic of the postmortem pathology of AD-afflicted brains. In simple terms, everything looked very good.

Readers should be appreciatively aware that the study of AD was significantly and substantially advanced as a direct result of a group of several hundred nuns who donated their brains for study of the amyloid plaques in AD. It is obviously difficult to use the approach of tissue sampling for any brain diseases, unless one settles for brain autopsy samples. These are important sources of information about the effects of chronic diseases, of chronic drug treatments, or of aging on the brain. This was an extraordinarily important study group in that their living conditions and diets were very regular, and, by being a close-knit community, they could watch and monitor each other and report accurately on cognitive or other decline. The authors would like to pay homage to these postmortem volunteers and remind readers that without volunteers there will be no new drugs.

The better, safer approach to vaccination is predictably mostly through "passive immunization"—making human antibodies in a fermenter and adding them slowly and stopping when the undesired autoimmune response occurs. However, though easier to control, it is more expensive and less attractive because at least monthly or quarterly injections are needed in cases where slowing disease progression is paramount. Whether making vaccines for Alzheimer disease or against tooth decay causing periodontitis, from a safety point of view the advice is to go for passive immunization because the adding of new antibodies can always be stopped.

Finally, whatever approach to new drug discovery one takes, be careful. Thalidomide was such a disaster in Europe that it is hard to believe it could make a comeback, but it has, which only goes to show that a truly efficacious drug is never dead. Thalidomide was approved in 1998 for treating a side effect of Hansen disease, commonly known as leprosy. The side effect is erythema nodosum leprosum (ENL), a debilitating and disfiguring lesion. The approval notice went on: "because of thalidomide's potential for causing birth defects, FDA invoked unprecedented regulatory authority to tightly control the marketing of thalidomide in the United States. A System for Thalidomide Education and Prescribing Safety (S.T.E.P.S.) oversight program has been initiated that includes limiting authorized prescribers and pharmacies, extensive patient education about the risks associated with thalidomide and a 100% patient registry. This oversight program is designed to help insure a zero tolerance policy for thalidomide exposure during pregnancy."[111] It is sad that we have not been able to make a better drug for leprosy, but did we really try?

[111] For more information about thalidomide (Thalomid) from Celgene Corporation of Warren, NJ, go to http://www.fda.gov/cder/news/thalinfo/default.htm

chapter

16

THE TRIBULATIONS OF CLINICAL TRIALS

WHAT WE SHOULD LEARN FROM FAILED TRIALS

What does one learn from a failed trial, and do we learn as much as we could? Trial samples and data on the effects of drugs—on successful and failed targets—still have value. Again, the recently proposed trial register will be a huge step in knowing what has been tried in man, both successfully and unsuccessfully.

The value of tissue and serum samples from trials that failed as well as from trials that succeeded is extremely important. This importance is now being recognized. Several drugs are "saved" from being stopped as ineffective or unsafe when analysis of the patients genetic makeup—their specific polymorphisms—of drug-metabolizing enzymes shows that all or most of those who had serious side effects belonged to a specific genotypically easily identifiable group that could in the future be screened. So, the FDA will insert a label saying: "Use of this drug should require establishment of a given polymorphism—from a blood sample—and carriers of this or that isoform of the drug-metabolizing enzyme are not to be given this medication." The principle is that there are no SAFE drugs in an absolute meaning, but drugs for which one can determine before treatment who might be at risk of side effects are much safer! The company will prefer not to lose the drug for which it has now fully paid the preclinical and clinical development costs and, more than likely, some of the premarketing budget. The market will sort out the rest: namely, if there is a drug with the same efficacy which does not require a blood sample, doctors will choose it first; if the drug has unique benefits, it will stay. Not all established drugs are easy to administer. Clozapine has been around as one of the unbeatable atypical antipsychotics, despite the need for continuous laboratory monitoring of serum samples; it simply has properties not found in any safer, more modern antipsychotic drug despite 40 years of trying to copy it.

Another example is the recent effort by GlaxoSmithKline (GSK) to find some distinguishing genotype—now found in the form of 89 SNPs ("Snips")—that would predict adverse effects of their HIV-antiviral drug (Abacovir) and enable patient screening, thereby permitting GSK to keep the drug on the market, at least until someone displaces it with a drug that does not require such screening efforts.

Companies are now storing millions of biological samples from trials as they have realized that these may help them to understand side effects and to limit the drug's use to those that will benefit, make it safer, and keep it on the market. What else they learn from these biological samples, and who owns that knowledge legally and morally, may be less certain. Companies that sell clinical samples and data are an important part of drug discovery, insurance rate determinations, hospital design, and so on and hopefully retain legal and ethical ownership of the human samples sold.

Agree in Advance with the FDA as to What Constitutes Success

The big thing is to agree with the FDA as to what would constitute efficacy and a successful trial. This is not simple because the outcome, apart from in infectious diseases, and from surgical procedures that fully remove a tumor, is never a "cure." Symptom treatment is easier than "disease progression modification." A 20% improvement in the patients' condition or slowing of the disease progression is often good enough to fulfill a criterion of efficacy, but sometimes this is considered not good enough by the market.

Trials of Disease Progression

If you want to look at the disease progression of, say, a neurodegenerative disease such as Alzheimer (AD), the length of the study becomes a major financial and marketing issue. Since the disease can now be detected much earlier, one can elect to perform the trial on patients with mild to moderate symptoms, as defined neuropsychologically, and try to show efficacy against a ***slow*** decline, but that can take 24 to 36 months. The market size will of course be much bigger; they are younger, and there are more of these patients who are less likely to die from other causes during the trials.[112] The difficulty is that the slope of decline in faculties is very shallow; thus, the difference between drug effect and placebo can only be seen after an extended period. If one chooses to perform the trial on more severe patients, where the decline is steeper, one can do a shorter, 12–24 month, trial, but here the risk of losing patients is greater since patients can die from other reasons, including natural causes. It is also more difficult to treat diseases once symptoms have become severe and what may work in moderate or mild cases may not work in severe cases. These—unassociated mortality and irreversible symptoms—are both problems shared in diseases with a late onset. So the company may have to square off market size and ease of gaining approval against financial risk and against later market entry—being number 4 rather number 1—although with a different, better label. Pharmacoeconomics and the life cycle of drugs management are key functions in successful drug companies, and they can generate more money by good decisions than discovery. Even these companies need some discoveries of new drugs occasionally, although they are quite willing to buy or license in rather than wait for the homegrown product.

The length of the trial, the statistical difficulty in establishing efficacy, even when aiming for only a slowing of progression by 20%, and the high dropout rate (up to 40%, would be a reasonable estimate) means that Pharma companies have to plan to recruit around 10–40,000 into a serious trial for AD. All of these patient/recruits would have to be seen by outside neurologists three times per year. Why "outside neurologists"? It is to avoid the potential criticism that might be levied if the experts were judged to be too close to the company. The FDA decided that for many things where the trial does not have a quantitative measure and where there is *judgment* involved in the "estimate," the trial needs to engage an outside expert who cannot be employed by the same hospital. And all of this tells you something about the cost of trial! The neurologists do not do this altruistically.

[112] The recent approval of the European drug memantine was for "mild to moderate" AD; all others—Aricept, Excellon, Remynyl—are for "moderate to severe" AD.

Marketing Moguls

Scientists are notoriously bad at marketing. Many scientists think that commercial utilization, or, rather as they might put it, "exploitation of science" is unseemly. Science (= *knowledge*) is self-contained and self-sufficient, and communication to peers is all that is strictly necessary for scientists to advance in academia. Scientists' own scientific peers determine what research will be funded. This may already be a flawed system since it is clear that the public, which, through taxes, pays for most research, wants more scientific progress especially in the area of curing disease. It is hard to argue against the logic. For example, many billions of dollars were spent over many years through the National Cancer Institute (NCI) on basic cell biology research without any cancers being cured. There was no point in attempting to answer questions about cancers without having more knowledge about cell biology. The logic of this was fully correct but it was pitched to Congress and the public as "Cancer Research." No one disputes the great progress that cancer treatment has made, but no one can argue that there is any cure for all cancer forms in sight. But perhaps it is too facile to portray scientists as impractical and nonapplied even when they are working on questions fundamental to understanding and treating disease.

Out of sight of the public and peers, the ambitious scientist is attracted to Pharma with the prospect of huge labs and huge resources. Whom do they meet? The marketing mogul!

Since this section of the book is largely for scientists and the authors are scientists, perhaps the authors should be overtly sympathetic. But to the marketing gurus poor communication from the R&D scientists is damaging. They can object to an entire drug development program if the preclinical research group can never explain simply, yet adequately, to marketing people or to the clinicians why this planned, new drug is different from the others. How is the drug going to give better therapeutic response, and how will it be easier to administer? The new mechanism *per se* and the scientific elegance do not impress them as doctors and patients do not buy new mechanisms; rather, they buy improved treatment, greater efficacy, fewer or less severe side effects, and easier administration. Of course, a new drug based on a new mechanism always gives hope and is often tested in treatment-resistant cases. Failure by the scientists to explain their case alone can lead to marketing not supporting the drug. It is purely poor communication within the company. Smart scientists can make it likely—as it is indeed likely—that a new mechanism of action or new drug target may translate into efficacy in patients resistant to today's treatments, and thus may open new markets. They have to come with good examples, and there are several, but they are not taught in "Pharmacology 101."[113] And imagine if the project leader is a medicinal chemist—they are definitely not taught in "Organic Chemistry 101." After years in a Pharma company, people learn all the key words of researchers, clinicians, and marketers and use them sometimes too often.

The R&D scientists should also listen very carefully to the marketing people. Drug discovery is a business, and marketing staff can recognize new openings in the market for drugs, either because a competitor's drug has been withdrawn or a drug that they thought would enter the market before their own candidate failed. So programs that were stopped because it seemed that they were too late in terms of projected market entry may get revived if the patents are not too old. However, the R&D scientists would be concerned that their own drug, which is more likely to be aimed at the same target and would be of a similar class, might fail for the same reasons as that of competitors. Sometimes one believes or wants to believe that the target is right; it is just that the other company had the wrong molecule. The argument goes that the competitor's program confirmed a great clinical Proof of Principle for us, but we have the right molecule. This is luckily true many times, and in Pharma a very convincing argument to start or revive projects is that the competition believes in them. The company would share their "risk aversion." More drug candidates are "killed" inside of Pharma than outside.

Standing Alone against Placebo

What is a "stand-alone drug"? It is one where you can seek approval for the drug to be licensed for administration with no other drug. This is important since, as we have mentioned before, it means you do not have to perform a more complex trial using your drug in concert or in competition with another. In reality you have to recognize that in no area of medicine is a single drug used to treat a complex disease, if for no other reason than wanting to exploit several mechanisms to achieve the same therapeutic effect, rather than having a single mechanism maximally exploited. For example, to lower blood pressure, one often combines a beta-blocker, a calcium channel blocker, and an antidiuretic. In addition, people with advanced age and high blood pressure often have high cholesterol levels and thus they are also on cholesterol-lowering drugs too. Why is this important for the team working on a new antihypertensive agent? First

[113] Within the United States, entry-level basic courses are commonly named "101."

because they have to test for drug interactions with all of these kinds of drugs specifically, or they may have to withdraw their drug very late, after $1 billion was spent, and the multibillion dollars in expected revenues vanishes.[114] Second, because one distinguishing feature may be precisely the lack of drug interactions.

Another area where one can be sure that no drug is used alone is in the treatment of pain, where our best drugs give only partial relief, and so drugs must be combined. In addition, many pain states have inflammatory components so that anti-inflammatory drugs are almost always included in the menu, and interaction of a new pain medicine with anti-inflammatories is not permissible.

[114] Pozicor, a novel highly effective antihypertensive drug, had interaction with the cholesterol-lowering statins and had to be withdrawn.

chapter

17

LINKING PUTATIVE TARGETS TO DISEASE STATES

CLEVER WAYS TO PHENOTYPE INCLUDING SELF-REPORTED PHENOTYPING

What information can genomics provide to the industry about human diseases? It provides validation of information that is core to decision making within the drug development process, especially where the decisions are effectively ***choices***. The validation process which the Pharma companies ask for, or, rather demand, from genomics, is answers to the questions: is this the ***right target***, is there a ***clinical correlation***, what is the ***tissue distribution*** for this new gene, some ***pathological samples***, and, preferably, ***linkage studies***—preferably from more than one population. All this would mean the company is on the right track when it comes to addressing a disease.

Even with all the developments of molecular biology and molecular genetics, one should not forget the significant information that can be gleaned from good doctoring. Sick people come to the doctor and tell him about their pains and ailments. This would be a ***self-reported phenotype***.[115] Unlike in animal experiments, where you have to be very ingenious to discover a disease phenotype, humans complain. This may be a strong clue in itself, but a phenotype is still preferably diagnosed as it relates to a clinical chemical and/or another identifiable parameter. This phenotypic marker is secondarily the gateway to a disease marker used in clinical trials.

These data are all hard to obtain from human studies. The industry relies on animal models of diseases to validate their processes. How does or how should the industry approach multigenic diseases in animal models?

In animal models of disease, the phenotype or marker is key to success. Scientists have for some time now had the ability to specifically introduce mutations of a given gene. Much was learned from looking at mutants of the fruit fly (*Drosophila melanogaster*) with visible differences in these engineered phenotypes such as different eye color, legs in place of antennae, and unusual behaviors. In mice genetic engineering produced classic phenotypes such as the "weaver" and the "reeler." Today we have the ability to examine the function of each protein—and each gene that encodes it—one by one involved in a disease process. We know how to make a so-called gene knockout or overexpressor, with the result, respectively, that a protein is fully absent or that the protein is made in abundance. Several Biotech and Pharma companies have generated tens of thousands of these "transgenics." The most rewarding procedure is to start systematically deleting one-by-one the proteins that belong to a class of good drug targets, for example, G-protein coupled receptors (GPCRs). It was known that: blockade of beta-adrenergic GPCRs leads to a lowering of blood pressure; blockade of the dopaminergic (D_2) GPCR leads to lowering of psychotic symptoms; blockade of histaminergic (H_2) GPCR leads to reduced acid secretion in the stomach. So why not knock out all the 450 GPCRs one by one and look for what pharmacological effects one can get? When the peptidergic (NK-1) type GPCR that binds the peptide substance P was knocked out, the animals had an altered pain threshold, with altered behavior in aggression and depression tests. Thus, a rationale for developing an NK-1 antagonist as an antidepressant as well as an analgesic was created; indeed, such NK-1 anatgonists are in clinical trials as a new type of antidepressant with the first proof of principle trials being in panic and post traumatic stress disorder (PTSD) both of which

[115] Phenotype: the outward appearance of the individual. It is the product of interactions between genes and between the genotype and the environment. There is now much talk about smaller *endotypes* but also visible, measurable changes.

present opportunity for much shorter and cheaper trials. Such phenotypes are not "visible" as are the weaver mice that have a specific motor behavior. The scientists had to test these animals with their littermates without the knockout of the receptor in several dozens of behavioral assays to demonstrate that both the NK-1 receptor was missing and this has an effect indicating reduced response to pain in a pain behavior model.[116] Having determined and confirmed the NK-1 receptor's involvement in pain, the procedure is to develop an antagonist to the receptor to mimic the qualitative effect of knocking out the receptor.

If scientists do not have the human data, then phenotyping is the "new" key for animal studies. However, scientists do not have, and need to invent and develop, sensitive, and clever ways to phenotype animal models of disease. If you do not have extremely good, sensitive, and clever ways to look for the phenotypic result of a genetic variation, then you will miss it. Most of the biologists, who work with transgenics, and by using forward mutagenesis, put an enormous emphasis on how to do it technically. The molecular biology and molecular genetics are extremely elegant, but scientists spend quite little time comparing the physiology and behavior—the "readout"—of the genetic changes (i.e., the phenotyping itself). The emphasis is on high-throughput tests since the generation of transgenics is rapid in particular in the so-called forward mutagenesis programs that use random mutagenesis of sperm. High throughput works well if you look for a cellular change in circulating blood cells, but if your mutation affects a subtle behavioral threshold, which might be the key to a new treatment for, say, panic, then you may easily miss the "phenotype." There is an excellent book by Jackie Crawley entitled *What's Wrong with my Mouse*?"[117] which addresses the question of "how to find out or uncover what is wrong with your transgenic mouse." Genomics increases the need for behavioral pharmacologists and for automation of "clinical chemistry for animals" with a better analysis of genetic variants of both

---

[116] Of course, mice do not go and put their tail on a hot plate so that you can register how long it takes before they pull it back, and they do not inject their paw to cause inflammatory pain. Scientists must be subtle in their approach to studying pain in animals if only to avoid distortion of data with, for example, stress. Decreasing stress can be misinterpreted as reduced pain in a poorly designed experiment. It may be of interest to readers to know that pain and anxiety might be considered biochemically linked. Some of the research in this area is conducted on the guinea pig since human NK-1 receptors are closer to those in guinea pigs than mice or rats. In rats the experiment doesn't work at all. An established anxiety test is in guinea pig pups which vocalize when separated from their mother. In knockout experiments the NK-1R(-/-) animals have altered anxiety and pain thresholds. The experiments suggest that an NK-1R antagonist would work in humans on pain and anxiety.

[117] JN Crawley *What's Wrong with My Mouse? Behavioral Phenotyping of Transgenic and Knockout Mice*. John Wiley & Sons, 2000.

knockout mice, where a gene has been completely silenced, and knock-down or overexpressors where the gene has been down- or upregulated, respectively, compared with the wild-type.[118]

The molecular genetic techniques employed in creating transgenics have exploded in their prevalence and scientific influence. But for sensible, solid reasons, they are largely used for isolating and affecting the activity of one gene per animal model. Unfortunately, most diseases do not have a simple genetic profile and are polygenic. For the Pharma industry to use the approach effectively, the power of the approach has to be refined. In an ideal future it would be possible to match the genetic profile of a complex disease with specific genes overexpressed and others underexpressed or dysfunctional within an animal model. Academic research is still some way from this since the genetic profile of complex polygenic diseases is not yet known. There is a lot of thinking about how to approach it.

One of the key weaknesses today is not in the physical manipulation of genes and their products, the proteins, but in the bioinformatics—the science of establishing, recording, storing, and interpreting the multiplicity of various changes among the tens of thousands of products of human genes. Gene function can vary in relation to changing diet, changing time of the day, with infections and inflammation, and can be different for different cell types. There is no disease, no meal that affects only one gene product. But to sort out which of the thousand or so changes in which cell type are key, which are the initiating and driving changes and which are derivative changes, and which could serve as a therapeutic drug target, with early and robust action, is a key task for bioinformatics, and it is being solved very slowly.

Genomics- to Proteomics-Derived Putative Targets

The key is to link phenotype to genotype. Linkage studies can indicate the correlation between disease state and changes in the drug targets' concentration or activity. Transcriptional and proteomic analysis, where one can show changes in the expression of mRNA levels or protein gene product in a disease or disease model, has been most successful if you have a chronic disease, so that there is time for both transcriptional and

[118] "Wild-type" is the term used to denote the normal animal.

translational changes to occur and manifest their effects. As such it can be an approach where new targets might be found in chronic diseases of the periphery and in oncology because of the possibility of obtaining the necessary tissue samples. The chronic disease often leads to a full remodeling of the cells and of the tissue affected. Thus, if one can harvest tissue samples from, say, a chronic inflammation area such as a joint in RA, one can find that there are large changes in the relative amounts of proteins between healthy and chronically affected tissue. In a lucky case some proteins are only found in the disease state. These provide excellent drug targets as they provide no target in healthy tissue, thereby providing a specific action only in diseased tissue. Such an example is the enzyme COX-2 targeted by Celebrex and its follower Vioxx. Aspirin and ibuprophen were long used as anti-inflammatory agents, but they hit COX-2 and also COX-1a—a close relative of COX-2—that is expressed in the lining of a healthy stomach and protects it from acid-caused damage. Hence aspirin, when used chronically to combat inflammation and to inhibit COX-2, causes stomach bleeding in many patients while Vioxx and Celebrex only affect the COX-2 expressed in the inflamed joint; thus they represented, it was thought, much safer drugs in this regard. Of course, there are no 100% safe drugs in all conditions for all patients, and indeed it is the added specificity that is ultimately the reason for Vioxx's unacceptable side effects.

Box 17.1 Why Vioxx Turned Out to Be "Bad"

There are several important aspects of the "trial by media" of Vioxx (rofecoxib), and Merck, and the FDA which bear additional scrutiny. The "evidence" that Vioxx causes an increase in cardiovascular (CV) incidents was by no means clear from the outset. There is equally certainly no clear evidence that the researchers in Merck or the FDA in approving Vioxx were anything but professional and objective in their judgments. Merck's voluntary withdrawal of Vioxx might be applauded rather than taken as evidence of sinister activity. There is no "smoking gun." What is the case against Vioxx? Many of the first trials showed that the 5–8,000 patients treated with Vioxx had no increase in CV incidents compared with patients on placebo. One trial where Vioxx was compared with another NSAID patients may have enjoyed a CV benefit from the naproxen. Other trial data are confounded by patients also taking aspirin which can be "cardio-protective" and "gastro-destructive." The statisticians can legitimately challenge the

continued

Box 17.1 Why Vioxx Turned Out to Be "Bad"—*cont'd*

basis of some of the 'reinterpretations of data' that have been published. The important point, which led to Vioxx being approved in the first place, is that gastrointestinal (GI) incidents were reduced and, moreover, pain and inflammation were reduced in patients with osteoarthritis. Interestingly, most of the accounts of Vioxx focus on the negative side effects rather than the positive primary effect that is to reduce pain by inhibiting the inflammatory process.

One needs to balance the views of, for example, Dr. Garret A. FitzGerald found in various articles in, for example, the *New England Journal of Medicine* (*NEJM*)[1] with Merck's own review as proposed by the president of Merck Research Laboratories (MRL) at the time, Dr. Edward M. Scolnick.[2] The apparent increased risk of CV incident using Vioxx was recognized, and clearly indicated on the label. CV incident might include thrombotic stroke, acute myocardial infarction (AMI), and sudden cardiac death (SCD), and so on. Moreover, the medical literature from especially Dr. FitzGerald, a consultant to Merck among others, gave a credible reason to explain the risk: Vioxx's own increased specificity compared to the other coxibs and NSAIDs might be the underlying reason.

A working assumption is that an increased specificity imparts greater efficacy at lower doses and reduces side effects. The specificity issue for the coxibs comes from hypotheses about the role of the "inducible" enzyme COX-2 in inflammation of the sort that occurs in joints with arthritis. The related COX-1 is thought to play a protective role in the gastrointestinal and cardiovascular systems. The "regular" NSAIDs that can cause in some individuals serious GI problems inhibit both enzymes. A specific inhibitor of COX-2 should be better especially for these individuals because it leaves COX-1 alone, while stopping the inflammation. Pfizer's Celebrex (celecoxib) was the first with a much higher specificity for COX-2 over COX-1 than the NSAIDs, and Merck's Vioxx followed with even greater COX-2 specificity. The Vioxx data bore out the promise: GI incidents were down. The drug was approved for chronic treatment of osteoarthritis and acute treatment of period pains. The dose for chronic treatment was not to exceed 25 mg/day and the 50 mg maximum dose for acute pain should not exceed 5 days. The maximum dose for ibuprofen is in contrast 2,400 mg/day.

It might be speculated that a specific COX-2 inhibitor acting with a localized effect on tissues undergoing prostaglandin-mediated inflammation such as at arthritic joints is a very good idea. But, what about inhibiting COX-2 in other non-inflamed tissues? COX-2 in balance with COX-1 in blood vessels and heart

[1] See, for example, FitzGerald GA, Coxibs & Cardiovascular Disease, NEJM 351;17 October 21, 2004, pp1709–1711.

[2] http://www.merck.com/newsroom/vioxx_withdrawal/pdf/VIOXX_scientific_review.pdf

Box 17.1 Why Vioxx Turned Out to Be "Bad"—*cont'd*

endometrium might be of significant physiological benefit. Might the effect of Vioxx be related to its upsetting the balance of enzyme activity in these nontarget tissues? Quite possibly, but while the hypothesis is intriguing and appealing there is as yet insufficient actual compelling evidence. While such arguments and concerns against the pursuit of specificity were voiced in 1999, there was no direct evidence against Vioxx in particular or other coxibs in general. And as mentioned before these were warnings about the possible side effects from interfering with a balance of incompletely understood biochemical mechanisms, not about whether Vioxx would be efficacious in providing pain relief.

Adverse drug reactions (ADRs) are thought to be responsible for over 100,000 deaths in the United States. each year[3]. Nonfatal but serious ADRs may number over 2 million in just hospitalized patients. The FDA granted Vioxx a priority review because the drug had the potential to provide a 'significant therapeutic advantage over existing approved drugs due to fewer GI side effects including bleeding' as reported by Dr. Sandra Kweder, the deputy director of the Office of New Drugs Center for Drug Evaluation and Research of the U.S. FDA.[4] The evidence according to Dr. Kweder "showed a significantly lower risk of GI ulcers, a significant source of serious side effects such as bleeding and death, in comparison with ibuprofen." A pain reliever with an improved GI profile would be a significant addition to the medicine cabinet.

Alleged problems started emerging with doses greater of especially 50 mg/day and with chronic use. Following approval for osteoarthritis and subsequently rheumatoid arthritis (RA), Merck embarked on a series of trials to extend the approved use of Vioxx for AD, prostate cancer, and colon polyps. The last trial, the Adenomatous Polyp Prevention on Vioxx (APPROVe) study, revealed a significant increase in thromboembolic adverse events after 18 months of treatment on 25 mg/day Vioxx. The APPROVe trial was stopped and Merck withdrew Vioxx on this evidence, even though part of the deviation between the control group and the Vioxx group might have been due to a peculiarity in the control group where CV incidence decreased. The hope is that the group on Vioxx will have a decreased CV incidence after being taken off the drug.

The seriousness of the side effect should not be underestimated. The *New York Times* has estimated—as have the trial lawyers sending out free class-action lawsuit case-soliciting letters to those who have a prescription of Vioxx—that altogether as many as 28,400 died while on Vioxx, but this does not mean the Vioxx was the causative or contributing reason. Some of the patients who

[3] http://www.fda.gov/cder/drug/drugReactions/default.htm

[4] http://www.fda.gov/ola/2004/vioxx1118.html

continued

Box 17.1 Why Vioxx Turned Out to Be "Bad"—*cont'd*

took Vioxx were not well, were aging, and may have been at risk of CV incidents. It might be that lawyers will try to prove a causality between taking Vioxx and any actual CV incident and, in addition, try to argue that all people who have ever taken Vioxx have an increased risk and that such a risk has not dissipated simply by ceasing taking Vioxx, that the risk is irreversible. The authors have no position on these arguments. If you would like to read more about what lawyers think about this then http://yourlawyer.com/ and http://www.coreynahman.com/vioxx_lawsuit.html provide superlative doses of vituperative invective.

Was Vioxx the appropriate medication for all the patients prescribed it? Might the physicians' organs such as the *NEJM* and *The Lancet* be protecting their clientele by not addressing this question? We cannot possibly say, just observe that the Vioxx question was couched under headlines such as "Failing the Public Health—Rofecoxib, Merck, and the FDA."[5] The truth of the matter is that drug companies and society often benefit from courageous "off-label prescribing" of approved drugs. In pediatrics the issue is acute in that very few drugs are ever formally tested on children for fairly obvious reasons, yet pediatricians are obliged to use adult medicines in scaled down doses.

What is of possible great concern is the amount of "direct to consumer" advertising budget for Vioxx. The success of Celebrex—becoming a $3 billion drug after the "fastest (market) uptake" ever—undoubtedly encouraged the Vioxx "bandwagon." The reputed $70–80 million annual advertising budget—coupled presumably with patient demand—made Vioxx the #2 or 3 drug at Merck in terms of sales. The "line extension" of Vioxx by trying it in trials for other conditions is symptomatic of the industry. Merck revealed Vioxx's CV problem with one of these trials. The APPROVe study was quite small compared to other trials with 2,600 patients enrolled. It is quite possible that Vioxx—which one would hope was prescribed to millions of patients for good reason—might make a comeback. For pain patients at low CV risk and high GI risk Vioxx might yet be good news. Recognizing which patients should not be prescribed a drug and making sure they have the right dose for the right time is a collective collaborative battle. Pharmacogenomic profiling and new drug delivery systems may yet make risks negligible or tolerable. The adversarial and confrontational nature of especially the U.S. "recrimination" system does not benefit society in the long term.

[5] *NEJM* October 21, 2004.

From Transcriptional Profile to Drug in Four Years—Jules Verne and the Therapeutic Antibodies

Normally, it takes more than 10 years to develop a new medicine. Therefore, to go from a transcriptional profile of a pathological sample to a drug in four years sounds like fiction written by Jules Verne or Isaac Asimov.

However, therapeutic antibodies to selectively expressed, genomics- or proteomics-provided targets, such as antigens in tumors, enter clinical trials for many cancer forms very rapidly. One makes humanized or human antibodies, by different methods, which either carry something or act on their own. These are safe. They are often efficacious. But they are difficult to administer; you have to inject them. They work in small indications, and they are perfect for small companies. The first Biotech company to make money in San Diego, IDEC Pharmaceuticals, did just this quite successfully, well enough later to buy Biogen and move to Boston in 2003.

The most celebrated of these therapeutic antibodies is Herceptin, which is used to treat breast cancer. It is a humanized antibody to HER1, a growth factor receptor on tumor cells. It is also the first drug for which genotyping is required. Before this very expensive—and not entirely side effect free—antibody is used, patients are genotyped or biopsies are examined to see if they express the target of this therapeutic antigen. The efficacy in the patient group thus selected is much higher than it was when all patients with breast tumors were the target group. Efficacy and safety are improved by genotypic marker use. In a manner, this may be the future model for "individualized medicine."

So, therapeutic antibodies are a class of drug therapy in which transcriptional profiling and proteomics have provided the targets, which are then used as antigens.

Biologicals Are the "Low Hanging Fruit" of Genomics-Based Target Discovery

Biologicals are the "low hanging fruit" of genomics-based target discovery. For example, if disease association were found with tumor necrosis factor (TNF-α) or interleukin (IL-1) expression, then the procedure would be to test a TNF-α or IL-1 antagonist in that disease. However, since TNF-α and IL-1 can stimulate each other, the new marketing "battle" for rheumatoid arthritis (RA) is between Amgen-(Immunex)-Wyeth's Enbrel, Amgen's parallel IL-1 receptor antagonist, Kinneret, Johnson & Johnson's TNF-α antibody Remicaid, and now Abbott's TNF antibody, Humira. These will compete on efficacy and perhaps safety, but also on ease of administration, which ranges from intravenous administration in a hospital setting (Remicaid[119]) to at-home subcutaneous injection for Humira. Enbrel (and the others) have been licensed for use as a stand-alone drug or in concert with the established RA drug methotrexate. Although Enbrel was approved specifically for psoriatic arthritis, it has not surprisingly found much general use in rheumatoid arthritis. Since TNF has a recognized involvement in many diseases (see Table 17.1), similar therapies may find much wider use.

Table 17.1 Association between TNF-α Polymorphisms[120] and Human Diseases

HTLV-1 uveitis	Primary biliary cirrhosis
Chronic active hepatitis B and C	Alcoholic steatohepatitis
Cerebral malaria	Asthma
Leprosy	Pneumoconiosis
Systemic juvenile chronic arthritis	Pulmonary sarcoidosis
Rheumatoid arthritis (RA)	Cardiac sarcoidosis
Primary sclerosing cholangitis	Ulcerative colitis
Systemic lupus erythematosus	Crohn's disease
Psoriasis	Ankylosing spondylitis
Myasthenia gravis	Narcolepsy
Multiple sclerosis (MS)	Body fat content

It is quite extraordinary to realize the large number of diseases and disorders where a polymorphism in TNF is implicated. Rheumatoid arthritis (RA) being a huge, very underserved indication, has become the focus of great attention (see text for details).

[119] Moreover, currently the marketing Remicaid, because it has been "assessed" to require administration by physicians, and not nurses or auxiliaries, has a "hidden" cost of the physician's time (say, $400 in the United States). The drug itself has been "competitively priced" at about $120.

[120] *Polymorphism*: the regular and simultaneous occurrence in a single interbreeding population of two or more discontinuous genotypes. The concept includes differences in genotypes ranging in size from a single nucleotide site (polymorphism, single nucleotide) to large nucleotide sequence visible at a chromosomal level. (From Medical Dictionary On-line: http://www.online-medical-dictionary.org/)

18

More Ways to Look for Targets

Reverse Pharmacology

Reverse pharmacology has great possibilities in drug discovery. *Classical pharmacology* starts with an active molecule and then looks for its effects and the receptor or enzyme through which these effects are exerted. For example, acetylcholine (ACh) was known from Loewy's experiments and from other studies in the periphery as an active substance. The target must be, by definition, an acetylcholine receptor. You knew it must exist; you already had a name for it when you found it; and all you had to do was find it, localize it, and discover its physicochemical, biochemical, and pharmacological characteristics.

Of course, to start with you did not know that there are dozens of acetylcholine receptors. The first bioassay of the contraction of striated, voluntary muscle, where ACh is a good agonist, reveals a nicotinic acetylcholine receptor. This muscle receptor through which muscles are contracted is from the same family that binds nicotine in your brain. In contrast, an assay of smooth muscle that exerts gut contraction reveals muscarinic acetylcholine receptors, and relatives of these regulate memory processes in the brain. There are at least four different muscarinic and more than seven nicotinic receptors. To find this out, chemists needed to make more selective ligands acting as agonists and antagonists to these receptors. This is all part of good classical twentieth century pharmacological research. Selecting these various acetylcholine receptors as drug targets is increasingly rewarding. Currently, agonists to nicotinic α-7 receptor are in trials for treatments of Parkinson disease, for cognitive enhancement, and for schizophrenia. Muscarinic ACh receptor antagonists are used to relax muscles prior to surgery. Some muscarinic agonists enhance salivation and pancreatic secretion, and others enhance memory performance in Alzheimer disease patients, to name but a few applications.

In reverse pharmacology, you start with a potential target. The Human Genome Project revealed sequences that you deduce, by using bioinformatics and data-mining, are probably encoding receptors. You hope that this may be a new drug target and simultaneously hope no one else at your competitors is looking at the same sequence. The aim is to find any agonist, or any antagonist to it. Of course, it would be nice to know what the natural, endogenous agonists are, but it is not a "must." You will then use the stimulating agonist or inhibiting antagonist to establish the chemical Proof of Principle of the importance of this new receptor in, say, pain models if you found that the protein is localized in sensory nerves.

In order to do this research independently, you need your own chemical library. If you, working in Biotech, haven't got one, you have to know someone, a rich uncle in Big Pharma, who has! Or, alternatively, buy your chemistry company as many rich Biotechs did (Celera, Decode, Millennium, etc.). Your ability to do this successfully depends entirely on the depth and breadth of your chemical library. The same target yields different "hits" in different people's—or, rather, companies'—hands. If your chemical library produces high-affinity agonists and antagonists to the target, you have a much better chance of examining the pharmacological and putative therapeutic value of the target. Once you have a chemical Proof of Principle you can establish whether it is worth developing a clinical candidate.

Academic scientists would of course go for a biological Proof of Principle. They would immediately delete the gene (i.e., make a knockout transgenic animal) and look for a phenotype. A Pharma company would, sooner or later, license this transgenic animal from the academic scientists to know if the compound they have made is specific. It should not work at all in the knockout animal if the drug is 100% specific since the target is not expressed and does not exist or occur in the knockout animal. The only interactions that should be present will be those responsible for causing side effects. This is very useful.

A few recent, real examples will indicate some of the possibilities and potential of this approach.

MAKING THE MOST OF SIDE EFFECTS

The new targets of this example are the caspases.[121] While this is indeed a real example, maintaining confidentiality precludes us from giving all details.

As background, it is known that there are 10 enzymes, caspase 1 through 10, and 3 of them are recognized as potential drug targets. They are important enzymes in a process called *apoptosis*—programmed cell death—that occurs, for example, in stroke. In stroke, the lack of oxygen to the brain—called *ischemia*—causes a cascade of biochemical steps that leads to cell death. It was quite a scientific breakthrough to establish that a cell's dying was an active and programmed biochemical process, and, therefore, not inevitable if you could inhibit an early part of the process. Thus, the targets are the caspases, and the therapeutic value is measured in preventing cell death after stroke.

This is a partial account of what happened. At the time of this story, not all the caspases were known. People were looking at both caspase-A and caspase-B, where A & B are numbers from 1 to 10. One group of investigators had an inhibitor [#1] for caspase-A, but,

Table 18.1 Inhibiting Cell Death by Inhibiting Caspases

	Inhibitor 1	*Inhibitor 2*	*Inhibitor 3*	*Effects on ischemic cell death*
Caspase X	x	xxx	x	xxx
Caspase A	xxx	x	x	x
Caspase B	xx	xxx	x	xx

Caspases are protein-cleaving enzymes that are involved in initiating the cascade of events leading to programmed cell death ("apoptosis") after, for example, ischemic stroke. Inhibitors of these enzymes inhibit particular caspases, of which at least 10 are known, by differing degrees. The strength of the inhibition is indicated in the figure by the number of **x**s. **x** is a weak inhibition, **xx** is a moderate inhibition, and **xxx** is a strong inhibition. The effect on ischemic cell death is similarly coded (see text for details).

[121] *Caspases* are a family of intracellular cysteine endopeptidases. They play a key role in inflammation and mammalian apoptosis. They are divided into two classes based on the lengths of their N-terminal prodomains. Caspases-1,-2,-4,-5,-8, and -10 have long prodomains and -3,-6,-7,-9 have short prodomains. [from Medical Dictionary Online]

seemingly unfortunately, it also inhibited caspase-B. Another group had a good inhibitor [#2] for caspase B, which was much more selective and had a poor affinity for caspase-A. They also had a poor general antagonist, inhibitor [#3] that worked equally poorly on caspase-A and -B. Incidentally, and importantly, neither group was then concerned or worried about the inhibitors' effects on caspase-X, since caspase-X was unknown at that time. Unfortunately, the patients' cells, unconcerned with the lack of knowledge of scientists, still carried caspase-X, whether the scientists were aware of it or not. Therefore, the patients had the side effect of the inhibitors' effects on caspase-X, which, it turns out, is an extremely important site. The result of inhibiting caspase-X is stronger in counteracting the negative effects of ischemia than is either of the other 2 caspases, A and B.

Once caspase-X was identified during transcription profiling of the ischemic cell-death model, they realized that the inhibitory effect of inhibitor [#2] on ischemia came from its actions on caspase-X, not on caspase-B. By using this cocktail of nonselective compounds, one has validated the target. In other words, because we work generally on classes of targets, we have ligands that interact with many specific targets in the same class but by differing degrees. They might individually be "lousy" drugs because of their lack of specificity or affinity to one particular target, but when used in combination with other equally or similarly lousy compounds, they permit you, with a little bit of imagination, to "cut a pharmacological profile" by determining the desired specificity with nonspecific agents. Now, you can focus on your validated target caspase-X and make, in this case, a specific inhibitor.

A Lesson in Molecular Pharmacology: Making a Bandwagon

Another way of doing reverse pharmacology is seen in an example on 5-HT_6 receptors. There are many serotonin receptors, and, by definition, they are good drug targets because chemists know how to make selective, high-affinity compounds that bind to them, from both medicinal chemistry and pharmacological points of view. It is relatively easy to make compounds that interact with serotonin receptors by making compounds that have an indole moiety (as in serotonin itself) or something similar to it in three dimensions. Researchers found a new receptor by *data-mining*—looking at possible amino acid sequences derived from genomic sequences—and, based on its possible structure but still somewhat speculatively and assumptively, called it 5-HT_6 (i.e., the sixth serotonergic receptor in the class). One looks for the localization in the brain of this receptor using a technique called *in situ* hybridization (ISH). Nobody had ever seen the gene product—the actual receptor—there was only a *probable* sequence from a computer. If you know the sequence, you can make a *nucleotide probe* that will bind to this sequence in the messenger RNA (mRNA) whenever and wherever the mRNA occurs and, therefore, where the receptor encoded by it occurs naturally. The probe is made with a fluorescent tag, which enables you to visualize it in a microscope. It was found that the receptor, which up till now was only suspected and expected to exist, is present in a region of the brain called the hippocampus, important for memory function.[122] It was shown that if you destroyed, by lesioning, the cholinergic nerve terminals in the hippocampus, the receptor disappeared. Thus, it was concluded that the new receptor sits on the cholinergic nerve endings. The solid implication is that serotonin acting here via this receptor affects acetylcholine release, but it didn't tell us whether it enhances or inhibits this release. Prevailing opinion was that it would be beneficial pharmacologically to enhance the release of ACh in the hippocampus because other evidence pointed to the fact that that would be good for improving cognition. Therefore, they injected an *antisense oligonucleotide* created specifically to downregulate (i.e., make less of this receptor) and it turned out that this itself enhanced ACh release as measured by microdialysis. The conclusion was, therefore, that the receptor normally inhibited the release of ACh and that a good antagonist to this receptor would have ACh release and memory-enhancing effects.

Subsequent to this, a behavioral study, using the Morris watermaze[123] to measure the spatial memory of rats, showed that the antisense to the 5-HT_6 receptor also enhanced cognitive function in behaving mice/rats. A Proof of Principle for the 5-HT_6 receptor antagonist as a memory enhancer was thereby provided, enabling the call within the Pharma company for a chemical program to make a truly selective compound to this receptor in humans.

---

[122] "Hippocampus" is named after its horseshoe shape. From years of research it is known to be important in memory assimilation.

[123] Named after Richard Morris of Edinburgh who invented this experimental process in which rodents (rats or mice) are required to remember where an invisible platform is in the tank in which they are obliged to swim.

What does it all mean? The antisense oligonucleotide downregulates the receptor, reducing its effects. We cannot easily use this approach in a human system because the antisense oligonucleotides are not stable or easy to administer.[124] In order to mimic the effect pharmacologically, companies would have to produce an antagonist to the receptor. This approach of reverse pharmacology had discovered, or at least uncovered, a new target for the treatment of cognitive improvement. Five companies at least are working on it. The experimental paradigm of phenotypic analysis of a transgenic, knockout, or knockdown mouse model, where one is obliged to look carefully for the phenotypic result of the molecular intervention, is now being widely used to discover and validate other targets. The results will only be as good as the phenotyping (i.e., "What indeed is wrong with my mouse?").

Had they looked for an effect on an inflammatory response, they would have found none. Crucial to the success of this approach was the scientists' making and testing the hypothesis that a hippocampal receptor regulating ACh release may affect memory function in certain tests, and thence showing that it was indeed the case. Inversely, the establishment of this model permits the *in vivo* test of the molecules synthesized by the chemists.

G-Protein Coupled Receptors

G-Protein coupled receptors (GPCRs) are very important because they are very ***drugable***. Many drug classes—β-blockers, angiotensin-receptor antagonists, dopamine D_2 antagonists, and H_2 histamine antagonists—act on GPCRs and sell collectively for approximately $12 billion per year. There has always been a race to find drugs for these receptors. Now, an interesting consequence of the Human Genome Project (HGP) is that we know the actual number of GPCRs. Olfactory receptors also belong in this class, but they are of much greater importance to perfume makers than to drug makers.[125] With high-throughput biology, Novartis, Pfizer, Merck, and others have already cloned and expressed all the GPCRs that exist. But they will need scientists to sort out which may be clinically interesting, based on distribution and function. Companies are additionally

---

[124] Although some tumor medicines are expected to come from antisense oligonucleotides.

[125] Previously, they used to be the same company. For example, Roche owned Givadon, the largest fragrance company, until 2000.

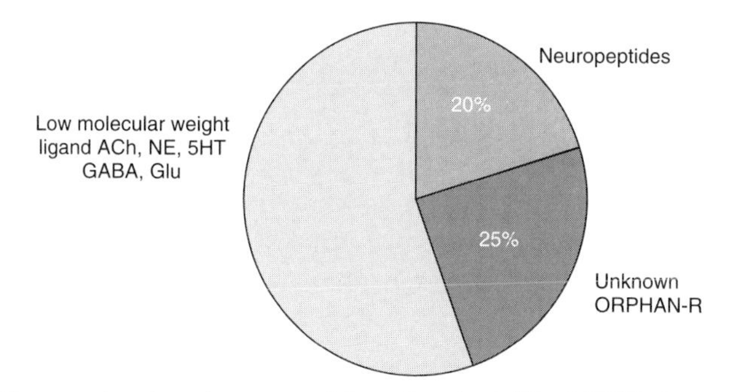

Over half of the receptors have known agonists or ligands. A fifth of GPCRs have neuropeptides as their agonists. Since GPCRs are very drugable, there is a race to find drugs targeting the remaining quarter of "orphan" receptors.

Figure 18.1 The 450 G-Protein Coupled Receptors in the Human Genome and Their Ligands

"keen" because their chemical libraries have hit rates of 0.5 to 1% for GPCRs, which denotes a high prospect of success in finding an initial clinical candidate upon which one can build.

Only 5% of the genome encodes for GCPRs; this translates as about 450 receptors. Already over half have known ligands such as acetylcholine (ACh), norepinephrine (NE), serotonin (5-HT), GABA, and glutamate. The specificity of action doesn't come from the uniqueness of the ligand but from the anatomical distribution and functional localization of the receptors. About a fifth of the GPCRs are known to be regulated by an army of neuropeptides. The remaining 25%—about a hundred—have as yet unknown natural or endogenous ligands.

We do know, however, that many of these receptors bind, for example, acetylcholine and serotonin. Serotonin alone is itself known to bind to at least eight receptors (5-HT-R_{1-8}). Therefore, we might speculate that there may only be another 10 to 50 endogenous ligands to discover and hundreds of pharmacologically important exogenous ligands: drugs to discover to the same receptors.

As is also seen from the two examples just given, however, we do not need to know the ligand in order to locate, characterize, and validate the receptors as targets. The companies don't have to find the endogenous ligands. They may be of major academic interest to someone who may want a Ph.D. in pharmacology or physiology, but the companies simply don't need to know the physiological agonist.

Orphan Receptors: Matching Drugs to Targets without Knowing Normal Function

Today you can find an agonist or an antagonist to any kind of receptor provided that you use a ***reporter system***. The clever thing here is that previously one had to look at radioactive labeling of something they were binding to the receptor, a very time-consuming thing to do. By using cells that can make a fluorescent protein (luciferase) in response to receptor activation (and cyclicAMP (cAMP) production), one changes the assay's readout from radiolabeling to optical labeling, and thus one is able to screen from 10,000 to 2,000,000 points a day. This generic method is brilliantly independent of the specific signal. The idea is you take a promiscuous G-protein that can act as a go-between for the receptor and the enzyme adenylate cyclase (AC). Then you test a large number of substances that may, via the receptor (i.e., a receptor agonist) and adenylate cyclase, activate the production of luciferase in the cell. The cell "lights up". The scientist who figured this out did so in the firefly, which uses luciferase in its nightly quest for mates; we are all grateful to her. This enables you to find agonists and antagonists and you don't have to know what the endogenous ligand is.

Once you have discovered the receptor's agonists or antagonists that are in your chemical library, you can go back *in vivo* to discover in which physiological control mechanism the receptor may be involved. As previously described above for the serotonergic receptor, antisense downregulation or transgenic knockout of the receptor suggests what the pharmacological effect of an antagonist drug would be. In addition, overexpression, by upregulating the receptor, may or may not suggest agonist-type effects. To test this, one would try to make a constitutively active mutant receptor to determine what the endogenous and any exogenous agonist may do pharmacologically.

The way you select which orphan receptor to work on first is that you look at the distribution of the receptor using a probe and *in situ* hybridization (ISH). If it is found in the hippocampus and you are interested in memory, you may start with this one. Homology to known GPCRs[126] may help determine function, but anatomical distribution strongly suggests function. If the receptor is sitting in the hypothalamus in the paraventricular nucleus (PVN), then—based on your knowledge of

[126] For example, if it is homologous to, say, the melanocortin-3 receptor, it would suggest an exploitable role that could really be worked on.

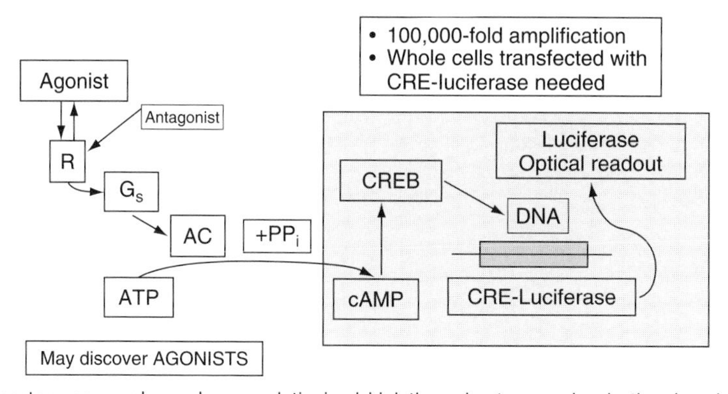

The reporter assay scheme has revolutionized high-throughput screening in the drug industry. Even if you do not know the endogenous agonist, one can test a million compounds a day against the receptor (R) belonging to the GPCR family. The G-protein (G_S) is coupled to adenylate cyclase (AC). If the candidate compound is an agonist to the receptor, the cell is stimulated through CREB and cyclic AMP (cAMP) which activates the luciferase and the cell is "lit up." The optical readout is detected automatically.

Figure 18.2 An Often Used Reporter Assay Scheme

functional neuroanatomy—you say: "Obviously, this may be a great target in obesity since feeding is regulated in this nucleus."

The major message is that screening for exogenous ligands is fully possible in the absence of knowledge of endogenous ligand. This was discussed in earlier chapters as it pertained to the morphine story, where use of morphine for thousands of years was not dependent on knowing the endogenous ligands leu- and met-enkephalin, and the endorphins.

We don't know what the endogenous ligands will be for the remaining orphan GPCRs. They may turn out to be neuropeptides, like hypocretin or nociceptin or then again, completely different classes. They may equally turn out to be modified fatty acids such as amandamide that binds to a well-known receptor: the cannabinoid receptor. Here again is a receptor that we know is producing mental and pain-relief effects since we have long used an exogenous ligand cannabinoid—as found in marijuana—to activate it. It worked without knowing that amandamide is the endogenous ligand. Before amandamide was found, who would have imagined it to be a signaling substance?

The endogenous agonists and antagonists may be very large molecules, say a peptide that is in specific cases, 11, 30, 36, or 41 amino acids long, or as large as trombin protein that also activates a GPCR just like the small molecules noradrenaline, histamine, and so on.

However, we might still expect to discover low-molecular-weight agonists and antagonists to act pharmacologically as orally administered drugs.

FUTURE TARGETS

What of the immediate future? We have already mentioned the reasons for optimism and will elaborate further as it relates to the business of the drug industry. In the next couple of years, the number of drug targets will grow from c.120 to 600 or so, but their distribution between target classes will be different, simply because the industry is obliged to expand its horizons while looking for new therapies and start to work on other types of targets. There may be an increase in the number of GPCRs from 75 to 120, but by 2010 it will only represent 20% of the drug targets, because the industry will have predictably redirected its attention to RNA-protein, DNA-protein, and protein–protein interaction drugs. By 2010 we should expect there to be 1,200 NCEs and 15,000 drugs. Hopefully, society will be willing to pay for the development and use of these drugs because they will have truly significant therapeutic effects.

part

III

The Business of Making and Selling Legal Drugs

chapter

19

THE BUSINESS BASICS (GENERAL)

INDUSTRY ATTRIBUTES

Now, this is veritable music to any investor's heart. As we stated earlier, the medicine market has increased continuously in the past 50 years by doubling every six years. Compare this to the car industry, or aerospace, or airlines, and it's starkly successful. This raises expectations in the market: a downturn would be decidedly embarrassing, yet Pfizer lost $4 billion in the third quarter of 2003 without too much alarm. Pfizer had just completed its second largest acquisition, and it was not a target of lawsuits, like the ones that brought down American Home products (Fen-Phen) or as was threatening the survival of Bayer (Baycol), so investors were not worried. How assured are continued profits and growth in the industry when there are now very real pressures on the industry from lobbyists, insurance companies, state and other governments, and, not least, patients demanding, for example, cheaper drugs? Is the continuous growth of drug companies a quite literal panacea? If drug companies fail to grow, will this expose the critical weakness of capitalism?

During these halcyon days the cost of drug discovery increased, but to compensate the companies grew. They grew organically, and then they merged, and they made acquisitions, so Glaxo became Glaxo-Wellcome, and then became GlaxoSmithKline. Astra and Zeneca made Astra-Zeneca. Ciba became Ciba-Geigy and merged with Sandoz to become Novartis to start the trend of renaming conglomerates and obscuring the new companies' origins. The French triplet Rhone-Poulenc-Rorer with the German Hoechst made Aventis, and so on and so forth. Perhaps it's not worth remembering them because tomorrow it will change. It did: the French-German Aventis allowed it to be taken over by the more French and rather smaller Sanofi-Synthélabo to make the world's third Biggest Pharma. Aventis had been resistant to being taken over by the Swiss Novartis.

The Pharma industry produces high-tech products at a high but somewhat surprisingly cost-effective price. The cumulative sales of the heartburn drug Losec are about $40 billion, but the savings for society are estimated at $85 billion because the drug reduced gastric surgeries as a result of gastric ulcer complications by 75%. But Losec is quite expensive at about $200,000/pound, which compares intriguingly with an F-18 Hornet aircraft that only costs about $400/pound.

How big is big? Pfizer, as measured by market capitalization value, may be, on a good day, bigger than General Electric (GE). GE, of course, is much larger in terms of number of employees, reinforcing the notion that the pharmaceutical industry enjoys a good earnings per employee ratio. Investors are becoming increasingly astute in their valuation of companies based on their actual projected future business (i.e., their drug discovery potential). Investors have to understand sentences like: "Modern drug discovery is about locating and characterizing a biological target, identifying 'hits' against this target with high-throughput screening (HTS) of chemical libraries, optimizing these hits, and selecting a clinical development candidate." They have to follow statements like "the FDA has given IND status" (or NDA status) to this drug candidate (or to a competing one), that "a second successful Phase III trial was conducted"—one needs three successful "pivotal trials"—and thus "the company has started to premarket this drug candidate," or that "the FDA proposed a new label" such as a warning label that "this drug may be carcinogenic," that is, effectively killing the joy over its great weight-reducing properties, and so on. But, as discussed in about 15 of the last 18 chapters, it is not always as easy as it sounds, and understanding that is important for anyone interested in the Pharma industry as a business.

Another important variable against which the industry should be measured is the product development time. Product cycle times in industries such as the car industry are measured in months, with consumers expecting annual improvements to whole ranges of a company's vehicles. In the Pharma industry the preclinical development time can be five to seven years, and the clinical discovery timeline is likely to be five to seven years on top of this. If the whole sequence of events goes smoothly, final patents on the drug are likely to be invoked during the lead optimization process, and, therefore, the patent-clock is ticking five to seven years before the drug launches. For a $1 billion per year drug, this represents a lifetime loss of revenue of around $5 to 7 billion or *$3 million per day.* Efficient clinical development is amply rewarded. Companies are only all too aware of this and push for "rapid followers," where they essentially cut out target selection and validation, and they start with high-throughput screening (HTS) to find another (patentable) molecule for the same target as has been just validated by a competitor's successful clinical trial or its entry into the market. So the exclusivity of drugs is a shrinking time scale. The time between the first beta-blocker and the second was six years; that between the first phosphodiesterase inhibitor (PDE5) and the second two and a half years. The first COX-2 inhibitor was followed by the second in just three months. And these are blockbuster drugs whose revenues are severely dented by competition.

Target validation requires a real biological innovation, and discovering an interesting new gene product: a new protein, or the so-called transcript that encodes it, and relating this to a physiological or pathophysiological state, is really rewarding. It may be a tiny part of drug development, but

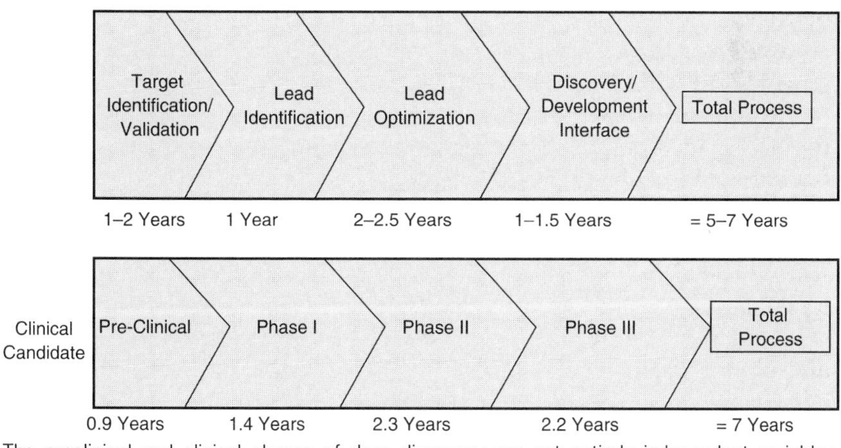

The preclinical and clinical phases of drug discovery are not entirely independent variables. There is overlap and interdependence in the process since preclinical work can extend beyond the Phase I trials and clinical development begins during the preclinical discovery.

Figure 19.1 Industry Averages of Preclinical and Clinical Discovery Timelines

these days, it is quintessential. The rest of the path to drug launch is long, 12 to 14 years, and arduous, significantly because drug development is **the most regulated human activity** under the purview of the FDA or its corresponding European and international counterparts.

But it is not so regulated or formulaic that anybody can do it. This is one of the great fallacies of small companies. Academic scientists who have made some great discovery either in the laboratory or from within a Biotech company, either try to form their own company or shop around for somebody to develop their idea for them. They often think it must be so valuable a finding that they expect a high price and are tempted to give it to the highest bidder. But they should realize that their lead may not be worth as much as they think, and they should definitely "give" it to a company that has developed another drug for the same indication, even if that company offers much less than a competitor. This is not only because 100% of half is a lot more than 100% of zero, but also because no matter how big a company might be, if the company has not previously entered this therapeutic area, they are more than likely to make a mess of it.

What exactly is the market for a drug and a drug company? At one level it is the number of patients who have adequate health insurance with the disease and in whom the disease is diagnosed properly; at another level it is, in the United States, the 90,000 general practitioners, or the 140,000 internists, or possibly only the 4,300 endocrinologists, or the 18,000 neurologists. It depends on who treats the indication.[127]

Another characteristic of the drug industry that makes it very attractive is the long shelf life of the products. Even when restricted at 15 to 20 years of patent, it is much longer than even a blockbuster from Hollywood or Bollywood. However, the dynamics of Biotech and Big Pharma could be likened to independent filmmakers. Biotech gets a talented team together, creates the products, hands them over to Big Pharma, playing the role of the Big Studios, and then the Biotech goes and works on the next project. Hollywood wouldn't like the Independents to get large enough to afford marketing, sales, and distribution, and neither would Big Pharma like to see Biotech be fully fledged.

Another characteristic shared by Big Pharma and Biotech is their need for intellectual property; they need patents! If you do not have them, your profits will be much lower. Companies in the United States can still

[127] The outrage felt in some quarters by Pharma companies marketing direct to physicians had plenty of time to fester. According to GSK's own history, in 1919 Mahlon Kline began the novel practice of sending pharmaceutical samples through the mail to doctors across the United States.

have an effective patent protection in some cases through the **orphan drug statute.**[128] But this is a very slow and difficult route; it's much better to patent discoveries.

Big Pharma Strengths

The most important single attribute holding an advantage for Big Pharma over Biotech is the internal expertise in clinical development. Expertise here is much harder to come by than in preclinical development. How many people in the world have developed a new antipsychotic? The clinical research organizations (CROs) may be a welcome help to Biotech, but experience shows that CROs will only do as good a job as your internal expertise can force them to do. The same CRO may do a lot better job for, say, Pfizer than for a Biotech on the same kind of trial. It will be more stringent as to which patients they admit to the trial and which clinical centers they use. On paper, it may seem equal, but it is only equal in the same way that all public schools are equal.

The other strength is a sales force. Most Biotechs do not have one, and Big Pharma thrives on them, especially in the States. The difference between having a good drug and making money out of it is the sales force. Every time a company wants to tell that it will make something new, it doesn't talk about its research. It talks about how many more thousands of representatives it hired who will visit physicians. This marketing machinery requires feeding with new drugs or formulations in the same area. In an extremely competitive marketplace, different drugs for the same target compete with each other and also with different drugs for different targets for the same disease. In other words, there are many ways to treat high blood pressure, and for each of these ways there are at least five different companies offering a drug.

Unlike Biotech, Pharma is importantly experienced in manufacturing, which is not as trivial as it sounds. It is all extremelly carefully regulated in the GMP (Good Manufacturing Practice) guidelines and followed up by FDA inspections. One needs to produce three identical batches of the drug to start with. One must not change the supplies of the simplest ingredient, etc. It is a complex issue of synthesis of exact compounds with exceptional purity in a form that is acceptable to the market. The Pharma industry starts with virtually valueless chemicals and creates a

[128] Under this statute companies are given a seven-year exclusivity. As an example, readers might refer to Dr. Ernie Beutler's discovery of 2-chloro-deoxy-adenosine.

prescription drug that is on average worth 24 times its weight in gold. It is perhaps the most extreme example of added value in any industry. Manufacturing facilities of even very big companies are often criticized after FDA inspections, for example, the Schering Plough factories in Mexico were closed in 2001.

On reflection, perhaps the obvious strength of Big Pharma over the rest is their ability, as a business, to "spread the risk." It does this by working in several therapeutic areas with drugs belonging to several target classes. With some drugs requiring small populations for short trials and others requiring 40,000 patients for two years or longer, if you can afford the whole spectrum and have many drugs in each category, you have balanced the risks of failures in the clinic. Biotech is often much better on the one, and, too often, the only target they have. This is a quite dangerous example of "putting all one's eggs in one basket" that can be disastrous in the business environment. If the Biotech's great idea fails in a Proof of Principle trial, it has nothing left. Both Big Pharma and Biotech often have the same response to deal with lack of product pipeline, and that is through acquisition. There have been recent examples where a Biotech company, whose major asset was its contracts with Big Pharma for services rendered, such as was the case for Millennium, has acquired another less successful Biotech that, in contrast, does have a product and pipeline. Big Pharma always has a revenue source from sales, whereas Biotech often has a revenue stream comprising contract agreements with Big or Medium Pharma. And Big Pharma in the past years has shown that it can also beat Biotechs on the biologicals market. Several of Big Pharma's best launches of late are typical Biotech-like products, but not licensed in; rather they are made at home to reduce the overall risk of only developing low molecular weight, orally available, classical medicines. At the same time, successful, wealthy Biotechs and smaller, poor Biotechs are struggling to acquire the know-how to make small molecule drugs and to have investors who have the endurance and capital to wait for five to six years for a proof of principle trial.

With all the advantages Big Pharma holds, it is intriguing to see in the current business environment that many drug companies are turning in a similar direction as Biotech, especially with regard to developing biologicals over traditional small molecules. Interestingly, in support of this observation the 2003 Biotech prize was shared by Novartis and Roche/Genentech for Xolair, an anti-IgE antibody. Roles between Biotech and Big Pharma may change as Big Pharma starts "to play it safe."

The deep pockets afforded by Big Pharma's history of profitable sales means that it is capable of throwing resources at an idea without needing clearance from financial overseers like nervous venture capitalists. Big Pharma can afford the expense, both in terms of funds and lost time, of a new trial simply to get a better label on its drug.[129]

Balanced Portfolio of Clinical Candidates

Big Pharma can afford to have, and indeed must have, a balanced portfolio of clinical candidates to fuel its product pipeline. And its active product line must in turn fuel its clinical candidate discovery and development programs. Clinical research is 5 to 20 times as expensive as preclinical research, and the potential liabilities in clinical research and in selling medicine are very large. But since the heads of clinical research at Big Pharma overbudget because of their expectation that some of the clinical candidates will fail, there is a much reduced chance of a catastrophic failure. Moreover, there is "upside potential" if more clinical candidates succeed. Of course, there is pressure on the preclinical R&D departments to prepare a significant number of these clinical candidates. The drug companies' large clinical research and regulatory departments assume steady flow of clinical candidates. They cope with the changes in volume by outsourcing some of the research (i.e., clinical trials) to clinical research organizations (CROs). The CROs help even out the flow when the business outlook has so many variables within its drug development programs. This makes for complicated management of the collection of programs.

The list of factors that come into play is indicative of the complexity. For example, the absolute risks of different trials ranges from no improvement while on the drug to actual increased mortality. The length of trials is different: antibiotics, 14 days, an Alzheimer drug, at least 24 months, and an osteoporosis drug is 36 to 48 months. The size of trials depends on the improvement one wishes to demonstrate and the natural rate of disease progression. One might only need 30 to 40 patients for a trial on some rare tumor disease, in stroke or sepsis, from 60 to 200 patients, but in cardiovascular medicine or obesity, one needs 2,000 to 10,000 patients to have some idea of efficacy. As a reminder, the key for these clinical trials is: "Is there a sufficient number of patients who are well characterized and who could enter the trial so that we will have the statistical power to

[129] This is done even at some staggering hundreds of millions of dollars in cost because calculations show that it is worth more to obtain a better or no label.

show efficacy, and also that the trial will not take forever?" In some cases, the best clinical sites and patient sources are already booked by the competition, which ironically may have already lost its drug candidate during toxicology testing. Many trials take a long time because the rate of enrollment is low. It is not uncommon that a 90-day drug trial takes 18 months to complete for all enrolled patients; it might be 90 days for each patient, but by the time the selected centers reached the required numbers $1^1/_2$ years have flown by.

To smooth out the bumps, curves, and reversals, each Big Pharma has 30 to 40 clinical trials ongoing all the time. So losing one is sad for the team that worked on it, but they can come back tomorrow to start on a new project. In Biotech the ups and downs are a real **drama** watched by investors and employees with the same anxiety!

Why So Many Biotechs?

With the many advantages enjoyed by Big Pharma, why are there so many Biotechs? It is rather like asking "Why are there so many rock bands?" It is something people want to do. The entrepreneurial band of scientists believes they can do better in developing their own product. Indeed, they have a point, as we have seen in earlier chapters about how vulnerable projects and even programs are in the hands of Big Pharma that do not need all projects to succeed. The radical projects one might find in Biotech do stand the best chance of success in the hands of their greatest protagonists: the scientists who conceived them.

But Biotech has weaknesses born from lack of experience, resources, and strength in depth and breadth. They also tend to try to do everything themselves rather than outsource them to CROs and even other Biotechs and because their investors dread the sharing of the expected profits. This is mostly because of lack of funds and partly because it takes specialists to do clever outsourcing. Lack of resources competes with lack of experience as the most important reason for this. Biotech doesn't seem to understand that without good support—toxicologists, clinicians, and volunteers—they might as well shut down the operation. How could Biotechs understand the subtleties of drug discovery? How many compounds did their executives and chief scientific officers put through the clinic? Two in their lifetime? Nobody is to blame, and for every 10—or 100—misguided Biotechs, there may be one or two great successes. The system and individuals are not to blame, but Biotech is a high-risk industry fueled by the dream of success of Big Pharma. Which will be successful? We cannot say! However, Fortune magazine in September 2005 (pp 250–268) calculated that c.$45 billion was lost by biotech investors since the start the start of biotech investments. It noted 157 companies listed on Nasdaq with a market captilization of $139 billion, while only 36 of the 157 had any revenue.

Alliances, Cooperations, and Partnerships

Alliances between Biotech and Pharma are well established and understood. There are new liaisons at the business level between different Biotechs partly in response to the poor prognosis for the economy in general and Biotechs in particular. So Amgen bought a 25% stake in Tularik and IDEC bought Biogen. The myriad of collaborations, alliances, cooperations, and partnerships are too numerous to catalogue, but to give an idea we reference an excellent source and resource, *Biocentury*, which in April 2002, catalogued Bayer's—i.e., a medium-sized Pharma company's—15 main collaborations shown in Table 19.1.

Table 19.1 Selected Collaborations with Bayer AG as of April 2002

Area	***Company***	
Bioinformatics*	Lion	
Combinatorial Chemistry	ArQule	
	ComGenex	
	Pharmacopeia	
Functional Genomics*	Affymetrix	Millennium
	CuraGen	Myriad
	Dyax	Nova DX
	Genome Therapies	Onyx
	Incyte	
High-Throughput Screening	CyBio	
	Novalon	
	Receptron	
Proteomics*	Oxford Asym	MorphoSys
	Genzyme	
	Genetics Inst	
Toxicogenomics	CuraGen	

A medium-sized Pharma has over 15 collaborations to identify and validate targets.

Source: BioCentury, The Bernstein Report on BioBusiness, p A3, April 29, 2002.

Other Players

As mentioned in Chapter 1, the Clinical Research Organisations (CROs) themselves are playing an increasingly important role in the business of Pharma and Biotech. They are becoming proactive in order to ensure their own pipeline of work.

chapter

20

ADDING VALUE IN A GROWTH INDUSTRY

R&D OF $26.4 BILLION PRODUCED NINE DRUGS

To recap part of chapter 8 (p 68), the efficiency of drug discovery is low. Like most businesses, it is easier to look at a snapshot of annual data than try to work out the historical costs of the business. Looking at 2001, which was not a particularly good or bad year, the total annual R&D budget of around $26.4 billion produced nine drugs—new chemical entities (NCEs). In addition, 26 biologicals—naturally occurring proteins (hormones and antibodies) used pharmacologically—were also NDAs: "New Drug Applications" that FDA approved. This number of NCEs is rather small compared with the 2,800 patent applications, approximately 200,000 compounds disclosed of the approximately 2 million compounds synthesized by medicinal chemists in drug companies. About 10% of the output from the medicinal chemists—making annually around 100 to 1,000 compounds per chemist—were considered worthy of patenting. Some of these compounds would have come from sources outside the drug companies themselves. "Library" companies (e.g., Alanex, Arqule, ChemBridge, Oxford Diversity, Discovery Partners) produce compounds on behalf of the drug companies and probably made around 3 to 10 million compounds. So, nine molecules (NCEs) "made it" out of 5 to 12 million made. Of course, the nine that made it had structures that were disclosed in patents filed between 1995 and 1998, when the number of compounds produced would have been less. But it is not clear that even though many more compounds are now being made "on demand," the output of NCEs will increase annually. It's a very complex relationship and a very long distance between the chemistry, the biology, the pharmacology, and the proving of safe and efficacious pharmaceutics leading to an approved new drug and to marketable medicine in the clinic with returns on the investment.

PATENTS

Summary information on patents is hard to come by. In most cases, the best intellectual property (IP) to own is the chemical entity that is the drug. Only some of the large companies would patent putative targets, that is, proteins, such as gene products, unless these proteins themselves will be drugs, as was the case for the '26' approved biologicals: erythropoietin (EPO), the TNF soluble receptor, and so on. For large companies even if they made some biologicals, the patent motto was: "We are in the business of making molecules; it's that, the molecule, that we want to hold the patent on not its target receptor." For small companies, their initial core IP might indeed be a gene or gene product and can be vitally important for financing and for assembling "work space." Many of their patents create a "freedom to operate"—and some others are truly to exclude or at least slow down competition. For example, corticotrophin releasing factor (CRF) is one of the most important hypothalamic hormones, involved in stress and depression. A small company, Neurocrine, took a patent on the CRF receptors, which it discovered, and wrote to all Big Pharma that they could not use the recombinant CRF receptor for screening when looking for a CRF receptor antagonist (for the indication of depression). What the Neurocrine patent does not prevent others using is naturally occurring, that is, not recombinantly expressed, CRF receptors from, for example, hypothalamic tumors. So, it is less facile to screen using these CRF receptor-rich tumors than using engineered cell lines with CRF receptors, but the patent restriction was mostly a nuisance for Big Pharma. Nevertheless it seems that Neurocrine has used its lead time well and was first in the clinic and made valuable partnerships with Big Pharma mostly in the wake of the interest in the CRF-antagonist program, which itself has seen its share of difficulties. But this is a common thread in Biotech: a talented group starts on Project 1 and gets capital—financial and intellectual—and finishes not Project 1 but Projects 2 and 3. But, eventually, the small companies still need to discover and patent the chemical entity that will bind to their gene product.

As we have stated previously, it is best to patent all your inventions. Businesses should patent not only the chemical intermediates and the chemical product, but also the method, (e.g., "method of inhibiting prostaglandin synthesis in a human host"), the formulation (e.g., pill, emulsion, capsule, powder), and the delivery mechanism (e.g., oral, intravenous (IV), intramuscular (im), etc.). But there is a strong cautionary note for Big Pharma as reported by the *Chemical & Engineering News* that speculated: "clashes over patent rights may hinder major drug producers' sales of some of the fastest growing and biggest selling pharmaceutical products."[130]

PRICE INSENSITIVITY AND OTHER BUSINESS DRIVERS

The drug industry as a business must ask itself: "How many D_2 dopamine antagonists, the major antipsychotic preparations, can the market bear?" The rule of thumb is that the first three drugs in any class do fine; they have traditionally done well. Pharma companies have historically never ever competed on price among drugs with patent protection. However, generics compete on price with all the patented drugs and with all other generics in their class. No two antipsychotics are sold against each other based on price. The companies try to profile them. The drugs have a different therapeutic ratio for different patients. Marketeers express this in the following way to the physician: "These drugs have these different side effect profiles, and these individual patients are differentially sensitive to these particular side effects, so you will figure out what's best for your patient." They may elaborate: "We're not going to lower the price because our drug has this particular side effect profile." How much would they discount a drug for a particular side effect? 20 cts for dryness of mouth, but nothing for sexual dysfunction? Or weight gain? They seem to conclude that they had better not take anything off the price.

The spectacular success of Lilly's olanzapine, which came years after Johnson & Johnson's risperidone and yet in the opinion of most psychiatrists it has the same profile, has taken everybody by surprise. First of all, all schizophrenics are diagnosed in all countries that can pay for the drug, and the disorder affects about 2% of the populations anywhere in the world, independent of race, culture, and economy. So, if everyone is diagnosed and treated, the only way the market can be generated for a new drug is if it is argued to doctors and patients to be so novel, and so

[130] http://www.reedsmith.com/library/publicationPrint.cfm?itemid=3825

much better, that physicians and insurers will buy it and leave cheaper drugs, most of which are equally effective in the treating of psychotic symptoms.

This requires ingenuity of research and marketing alike. Lilly succeeded in several of the U.S. states—most of which pay for this chronic drug treatment—to make, according to the *New York Times*, the antipsychotic olanzapine the single largest line item in the budget.

It would also be possible that one makes a drug, which addresses the nearly 30% treatment-resistant cases, but that probably would require a truly new drug target and a new mechanism of action. Such a drug, even if it only worked in the treatment-resistant cases, would sell for more than the largest antipsychotic on the market now since none has 50% market share of the remaining 70% of patients who are responders to D_2-antagonist-type drugs.

But it's becoming different for biologicals. Biologicals should be and usually are the same protein molecule as that which occurs naturally, such as insulin, growth hormone, tPA, interferon-α, and interferon-β. The patents for the gene or the transcript offers some protection, but later if not sooner, a competitor will come out with a new formulation for your biological, or a new patentable way of making it. Techniques or tactics might include modifying the structure to prolong activity and maintain efficacy, such as by the method of *pegylation*, or they can be put into slow release—*depot*—preparations, as was being done for EPO, so that you may now have an injection once a month, not every day. But since it is the same molecule and since differentiation by formulation can have only some limited value, biologicals are the first Pharma products where price competition started. Insulin, growth hormone, interferon-β, all started to compete on price, but even then there was a strong drive to compete on medical qualities. How can this be if the molecules are the same? Perhaps the manufacturing method produces some "aberrant" molecules, where, for example, an individual protein molecule hasn't folded correctly and is thus ineffective. As previously mentioned, Schering, Biogen,[131] and Serono all have interferon-β for the treatment of multiple sclerosis, and because Biogen's has to be dosed much lower, it is now saying that "You have to examine how much antibody response you get to this drug," and "Because we produce less antibody response, our treatment will be standing up for a longer period, and, therefore, it is better." They are trying to compete on efficacy, convenience, quality, and now "antigenicity" of their product. All of this is perfectly reasonable scientifically and medically. Just as

[131] Now IDEC Pharmaceuticals.

depot-preparations are not only for convenience but are medically important as the activity level of the protein in the body is more even over time than with repeated injections. Some of the reasons may not be very compelling, so they reluctantly have to recognize price as a factor in the physician's decision making. Insurance companies and governments are becoming more focused on this major issue. But the guiding principle of the Pharma industry remains: "We should not compete on price."

Many analysts openly state that there is "no price competition in biologicals" and "there is no roof of pricing for them." They quote the success of recently approved therapeutic antibodies to treat different types of cancer, some costing $20 to 30,000 per year, and their effect according to the manufacturers is a prolongation of life of two to three months. Whether society indeed will accept and can pay for the result of the presently ongoing 350 trials in oncology with therapeutic antibodies is highly questionable. Many of these treatments will be approved by the FDA over the next five years. If all were accepted into the marketplace, it could raise the medicine bill by 25%. We doubt that this will happen without much debate, and a ceiling on what the individual drugs may cost might be agreed on, if not imposed. And this will be without individual competition for the drugs but will be a result of competition for medical budget dollars.

As marketeers customarily and often argue and as the FDA requires, sometimes even the new drug is run during its trials against not only placebo, but against a positive comparator, a drug known to be efficacious in this condition. It does not have to have the same mechanism of action but it helps to know if the trial was properly run. Companies and regulators alike understand that one center of a multicenter trial in which even the positive comparator, which is known from clinical practice to be effective in this condition, did not work, possibly did not do a good job in selecting patients or in evaluating the response. So it is helpful. Why do we recount this here? Because no one likes to pick as comparator the BEST drug, a just-approved, good drug, even if in the marketplace you will run against the best drug "gold standard."

This is why it was so surprising when Pfizer sponsored a study to compare its long-marketed best-selling—$2 billion/year—cholesterol-lowering drug Lipitor, originally made by Warner-Lambert, against the more recently introduced cholesterol-lowering drug Parvachol of Bristol-Meyer-Squibb (BMS) that was gaining market share.[132] This is, of course, a Phase IV study to support marketing, not for finding a new indication as

[132] For one account, readers may consult the *New York Times* article in the November 16, 2004 edition.

have other Phase IV trials such as for Prozac (first approved to treat depression, then panic, and then social phobia) but to beat down emerging competition since Parvachol, when approved, was not running against Lipitor in its clinical trials.

Sometimes, the choice of comparator can backfire badly. Zocor Merck's Statin was run by Merck against Pfizer/Warner Lambert's Lipitor in another billion dollar study, and it was proven less effective. Merck made a huge favor to Pfizer that paid dearly but astutely to acquire Lipitor for its cardiovascular franchise.

Of course, the same price competition exists for generic drugs as for laundry detergents, and the competition to be first to gain approval for a generic drug is murderous. In addition, the former owner of a drug may and often stays in making the generic drug too, nowadays as Big Pharma has successful OTC and generic businesses (Novartis, for example, having bought the largest German generic drug maker Hexal in 2005).

Box 20.1 Why Vaccines Are a Bad or Not So Good Business

Vaccines are the oldest biologicals. They also have the largest impact on global health, alongside with clean drinking water. In view of these facts, it is astonishing to note that there are very few vaccine manufacturers in the world, and that some of the best, most historically important, and famous ones have been closed in the last two decades. The reason? They are not showing a great profit margin. "To sell vaccines is a low-margin, high-risk business," a Glaxo executive reportedly said in commenting on closing the vaccine manufacturing at Wellcome when Glaxo achieved control over Wellcome in 1995. Wellcome, and the "spawned" Wellcome Trust, was one of the most respected organizations in the world of health care, especially for antimicrobials and vaccines being the inventors of numerous vaccines used all over the world, including the rubella vaccine in 1971. The perceived lack of incentive for commercial development of vaccines has led to the reduced number of vaccine manufacturers. While vaccines are given to 400 million people per year and save possibly 40 to 200 million lives per year compared with an estimated 10 to 40 million saved by drugs, the market is measured in $10s of billions compared to almost $800 billion for the drugs market.

Much of the problem can be attributed to the fact that many of the diseases requiring vaccine development are acquired in the developing world. A new tuberculosis vaccine and any HIV vaccine would be welcome in the developed as well as the developing world, but a much needed malaria vaccine is only sought after for the developing world.

Private and public funding of foundations such as the Global Alliance for Vaccines and Immunization (GAVI), supported by the Bill & Melinda Gates

Box 20.1 Why Vaccines Are a Bad or Not So Good Business—*cont'd*

Foundation and the Government of Norway, among many others, is aimed primarily at bringing known and available vaccines against diphtheria, tetanus, pertussis (whooping cough), measles, tuberculosis, and polio to children in developing countries who are not being immunized. GAVI is also introducing underutilized vaccines such as hepatitis B, *Haemophilus influenzane* type b (Hib), and yellow fever. A stated ambition is to work with vaccine makers to ensure a reliable supply of lifesaving vaccines with a "guarantee of predictable, long-term markets," encouraging "greater competition—leading to reduced prices," and stimulating "additional investment in R&D for urgently needed vaccines." The Gates Foundation is committed to funding research initiatives to develop vaccines against, for example, HIV/AIDS, TB, polio, dengue and dengue hemorrhagic fever, and malaria.[133] Will this make a difference to the industry? Possibly. Big Pharma is unlikely to change strategy, but some Biotechs might benefit and develop vaccines backed by such funding.

Other groups and initiatives can provide commercial incentive. The development and manufacture of viral vaccines for medical staff and those exposed against hepatitis A and B was a great step in improving the safety of those health care professionals. Really substantial profits for manufacturers can result once a safe vaccine for a professional group is developed and its use spread. The margins of those vaccines are fully comparable to those of successful medicines. The race is on to develop hepatitis C vaccine, and as knowledge of hep C's role in liver cancer accumulates, success is more likely.

Tourism also provides a market for vaccines and antimicrobial agents. The names change with the locality, but the avoidance of diarrhea from local *E. coli* variants such as India's "Delhi-belly" or South America's "Montezuma's revenge" provides a market. We often forget that the largest organized tours of foreign land are military campaigns. Many more days are lost for troops because of diarrhea than because of enemy action. The largest buyer of the *E. coli*/cholera vaccine is the U.S. Army. Tourist vaccines command estimated prices of $30 to $500 per dose, and the U.S. Army reportedly pays $100 to $300 per vaccine per soldier.

A special aspect of vaccine development involves the need to keep a number of vaccinations low because of the problems in getting patients to a vaccination site, because of the risk of infection from the injection itself, and because of the need of an "adjuvant" that boosts the response of the immune system to the pathogen. Modern vaccines contain several components that will provide antibody response to several pathogens and thus protect against several diseases. In the popular triple vaccine (tetanus-diphtheria-pertussis), each component also

continued

[133] See for example: http://www.gatesfoundation.org/GlobalHealth/InfectiousDiseases/Vaccines/Announcements/Announce-050124.htm

Box 20.1 Why Vaccines Are a Bad or Not So Good Business—*cont'd*

serves as adjuvant for the other compound. There is, however, a limit to the number of antigens the immune system can take on at the same time, before getting exhausted. Thus, we are unlikely to see more than three to five pathogenic antigens combined in the same vaccine. *Vaccinia* virus is presently being developed for the delivery of multiple viral and bacterial antigens. The goal is to have a multiple vaccine with an 80 to 90% protection rate and for the cost of $3 or below.

The dilemma of vaccine development centers around the field trials: The associated ethical, political, and epidemiological problems are key to understanding the slow pace of vaccine development. The largest of all "obstacles" in developing new vaccines is the issue of litigation. For many vaccination programs that form the basis of public health, the government has to guarantee manufacturers support and protection against lawsuits. Enforcement of vaccination programs to the extent that the unvaccinated cohort does not get so large that new epidemics can start is essential. The refusal of vaccination on the part of some religious groups is in stark contrast to the heroic efforts of many African women who walk on average of six hours with their children to the vaccination site.

Are sufficient numbers of new vaccines being developed? HIV vaccine efforts have been ongoing for the best part of 20 years, and some small vaccine trials in different African countries have been carried out with variable results. This might be considered lucky for politicians, insurance companies, and bioethicists in the United States. Consider their dilemma if the vaccines against HIV looked a bit better. Assume that a new HIV vaccine is going to be as good as the polio vaccine. Namely 1 to 5 in 100,000 might have serious side effects, and some vaccinated subjects would get HIV/AIDS as a result of vaccination. Unlike polio, which had large epidemics, HIV infection is sexually transmitted. Whom would society vaccinate? All children? Sexually active people? Risk groups? Who would take care of those who get the disease as a result of vaccination and not as a result of "voluntary actions"? The checkered history in all countries of how HIV infection caused by contaminated blood supply was dealt with is not giving much hope for fair treatment of vaccination-caused disease cases—when the disease is associated with so much stigma.

Biodefense initiatives in the United States, Europe, and Russia include development of vaccines against pathogens that are suspected to be developed as biological weapons by adversaries. The development of biological weapons is prohibited by numerous international conventions (see redcross web site www.scienceforhumanity.org). This also means that a developer of the defensively oriented vaccine against, for example, "Venezuelan hemorrhagic fever" cannot legally test the effectiveness of this vaccine because it is forbidden to manufacture, store, or expose anyone to the virus as it is classed as a biological weapon. So either one vaccinates with an unproven vaccine and exposes one's own soldiers to risks without established benefits, or a convention is broken. In this context, many U.S. soldiers refused to be vaccinated against anthrax, which could be treated with antibiotics if needed.

Foreseen and Unforeseen Costs

The threat of litigation makes drug companies err on the side of caution. Many drugs are dropped before they reach the market; others are withdrawn after they have started to generate revenues, some significant. The seven drugs that have been withdrawn in the last three years were selling for $11 billion per year, and represent huge losses for those companies. Because the withdrawals were all from interactions with other drugs, these drugs were not unsafe when tested by themselves. Baycol (cerivastatin), a cholesterol-lowering statin, was withdrawn voluntarily by Bayer because it was found to be potentially fatal in combination with another drug, gemfibrozil.[134] Bayer had originally warned against using it with gemfibrozil, and it also recommended starting patients on a low dose and building up to the maximum dose of Baycol. In a number of lawsuits, Bayer has in fact been exonerated in some courts because the drugs were prescribed against the instructions on the label. Baycol was prescribed by physicians at maximum dose to their patients already on gemfibrozil. But that doesn't mean that Bayer will not have to continue to defend itself against lawsuits in a number of other U.S. states. At the time of writing, in California at least, lawyers were still advertising on the television for patients who were prescribed Baycol to come forward to join class-action suits. The lawyers are not interested in suing the doctors; they aim at the whole company, whoever has the deeper pockets. A Big Pharma is not sued for the damages it causes with a particular product but for the worth of the company, for the worth of all its products. Sued companies like American Home Products (now part of Wyeth), Bayer, and Merck have to make substantial disposition of capital to be able to pay damages that may be awarded and as long as there are pending claims, not only in the form of class-action suits, but also by individuals who did not agree to be represented in a class-action suit. This is not trivial—no one knows the potential full cost—and by many standards is not fair. In other industries, it is not like that. If you make a tire that doesn't work, you have to recall just those cars fitted with that tire and change the tire usually with no compensation unless people were hurt in an accident. In contrast, the way asbestos lawsuits now wreck whole companies is not unlike what happens or what might happen to some Pharma.

Of course, as long as there are suits against you, no one does business with you, you cannot be bought or merged, and you have to settle. It is a perilous position for a company that is set up to help physicians treat

[134] See, for example, http://www.fda.gov/cder/drug/infopage/baycol/baycol-qa.htm or (commercial) http://www.baycol-law.com

patients. Inappropriate use of drugs is behind almost every withdrawal. The rest comes from overzealous marketing forces not listening to the companies' own safety warnings.

A more predictable expense for a company is drug failures. Every significant company expects some of its candidates to fail and budgets accordingly. But the costs are driven up if the failures are late in Phase III or beyond. Better to fail early, which makes it very tempting and easy for companies to "pull the plug" on projects prematurely. This is behind the reason for the price per earning for Pharma stocks being so different from car company stocks. It is because we really cannot predict very well; there are great upsides and great downturns.

This (see Figure 20.1) is a very key figure for the industry. It shows that projects and candidate drugs fail. In 1997, the attrition rate was such that in preclinical research 50% of the projects did not lead to a drug candidate

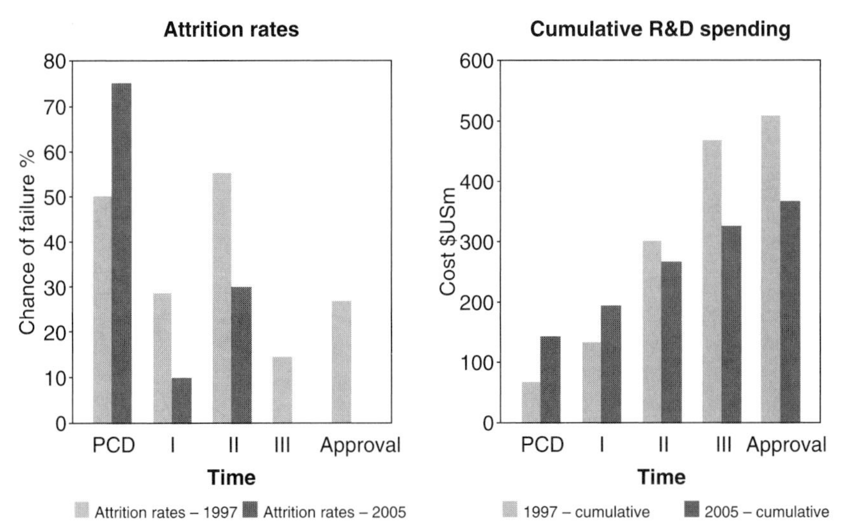

Drugs fail, and when they fail makes a huge difference in the financial costs. On the left, the attrition rates (chance of failure) at stages of preclinical development (PCD), Phase I (I), Phase II (II), Phase III (III), and when approved ("Approval") for 1997 (lighter) and projected for 2005 (darker). On the right, the cumulative R&D spend for the same two periods. The new slogan for clinical development is: "Fail early only start clinical trials if you have a biomarker to follow the drug's effects." The biomarker may not have to be approved by the FDA, but needs to be just good enough to convince the company that the drug is doing something positive. Figure originally from Price Waterhouse Cooper.

Figure 20.1 Costs Are Driven Up by the Late Failures

worth putting into clinical trials. In Phase I when the companies look at safety in the drugs that they believe are efficacious and worth pursuing, a further 30% "die" despite extensive toxicological studies in three animal species before Phase I. Of those drug candidates that survive, another 55% die in Phase II and a further 15% of them are still dying in Phase III. One hundred clinical candidates have become 13 launched drugs. But the most fiscally painful time to die is once they are out on the market when you will lose another 25% and be down to 10 out of your original 100 clinical candidates, and the original 200 preclinical research projects. And when they die post-approval, that is when the lawyers send out the letters which say: "Dear Neighbor: If you had been on Baycol we would love to represent you. Let's sue Bayer together." This is really not great for the company, for the industry, and, ultimately, for society. There's no such thing as a free lawsuit.

The industry would like to do better. By 2005, the industry would like to get so smart that it only fails early because the cost of failing in the preclinical (PCD) is one hundredth of failing in Phase III and one millionth of failing after approval. Cumulatively, the cost of failure should go down. The cost today has risen since 1997, and now it may be $800 million development cost per successful drug. Without trying to be discouraging about the drug industry as a business, the two influenza drugs launched in 2001 altogether only made $80 million while costing some $300 million a piece in manufacturing, sales, and marketing before Tamiflu was rescued by the stockpiling against the threat of a flu pandemic. The initial poor performance against expectations was solely the fault of marketers who pushed these drugs forward at any cost and made the wrong assumptions that vaccines will lose market share and that there will be equally big flu epidemics each year. Well, vaccines are improving: the first nasal flu vaccine is out if you are afraid of the needle; you can get your vaccine the way you will get your virus when someone infects you—through your nose—and you do not have to pay through the nose; it is relatively cheap. That marketers complained that there was not a big enough epidemic is a weak defense and speaks volumes about their performances. In other cases such as for Lilly's olanzapine the market was underestimated by a factor of 3–5.

Incidentally, Big Pharma cannot easily license out compounds, about which it is unsure, to smaller companies because the original company will still be sued if things go wrong. They don't make "paper barriers"[135] strong enough to protect you; you have to kill any drug that is in doubt. Now some companies pick up failed projects at Big Pharma and buy huge insurances, but it is unclear if it indeed will protect Big Pharma in the end. We have not

[135] In this context "paper barriers" refer to legal distancing of yourself and the drug.

yet seen this because no drug that they have finally, in these cash-strapped times, licensed out to a small company has been approved, and, thereafter, had such bad side effects that lawsuits had started. But these are the early days in the history of Big Pharma licensing out projects and compounds. Ten years ago the purpose of the licensing department was solely "in-licensing." Bonus pay was only given for in-licensing. The majority of Pfizer's drugs have come from in-licensing or acquisition. The prevailing culture did not encourage spending time on "out-licensing" because no potential buyer believed that something Pfizer could not do with its experience was either doable or worth doing. The expected rewards were projected to be small. Virtually nothing up front would come from the small companies since they do not have the cash. Then came the mergers, and Big Pharma often found themselves with two or more development projects for the same indication even for the same target. They could now make the argument that while both might be indeed very good drug candidates it would be better to license one to a smaller company. They might even license the more promising candidate. Of course, they may retain future marketing and sales rights. In addition, directors who choose or who are "invited" to leave often know the virtue of a drug with which the merged company would have internal competition. A few very successful examples of out-licensed drugs are appearing. Octillion is the biggest Biotech in Switzerland—some CHF6 billion on the Swiss stock market—primarily because of a drug its founders licensed from Roche and on which they worked at Roche for 10 years previously.

What would make you decide to stop clinical development? Sometimes marketing, sometimes your own evidence, but sometimes just because other companies' trials for the same target failed. Unfortunately, the other company may have had a poor compound, or it may have been "frightened" off its drug for a variety of miscellaneous reasons. But the threat of litigation makes saying "stop" easier than "continue" or "press on regardless." The rumors at scientific meetings that involve clinicians who do trials for Big Pharma can be lethal for any drug candidate in development. There is enormous potential value for patients and companies that is destroyed by rumors at these meetings by people who cannot be quoted, cannot be held responsible, and have nothing to lose. One needs to be a strong executive to stay the course when five colleagues come back with the "top secret" info gleaned at expensive dinners from delegates with hopes of a rich consultancy that the competitor's drug has serious problems, or was stopped for "possibly mechanism-related reasons." To stay the course is hard, especially when you have budgeted for failures, and since you cannot sponsor every trial you started. The rumor has a very strong voice in stopping—many times unnecessarily—drugs that could have been good for patients and companies. One might wish that scientists and clinicians, who at meetings often without seeing all the trial data "gossip" about how good, robust, or bad a drug is, should be held responsible, if not actually liable, for the rumors, as they are held responsible for the outcome of spending their research grants, which are a thousand times smaller than the values they destroy with ill-considered speculative opinion.

Landscape Perturbations

The business landscape for the pharmaceutical industry is often changing, or being perturbed at least. All the best plans may fall foul of competition, both direct—when your competitor produces a similar drug or its drug becomes generic—or indirect—when your competitor approaches the problem in such a different way that the market no longer needs your drug, such as might be the case if an anti-angiogenesis drug were more effective than any solid tumor drug. Changes in the landscape stop drug development as often as do scientific concerns.

There are also possible changes in society or the population at large that can make life difficult for the Pharma industry. Nowadays it may not be possible to vaccinate the general population with a vaccine such as smallpox, which is a live vaccine, because if the patient were immunocompromized in any way, then the vaccine would be fatal. This would include anyone who had had an organ transplant and those with a chronic inflammatory disease (and receiving immunosuppressive drugs such as methotrexate). These would be the largest groups in the estimated 18 million immunocompromized people in the United States. Even those who self-medicate on cortisone creams may be immunocompromised, but there are no accurate data on absorption through the skin, so we have no idea if they are. Of course, HIV infection is a significant, though much smaller, risk factor with 200,000 to 2 million affected patients with immunosuppression caused by HIV itself and not by a drug. The U.S. government worried about bioterrorism may have bought all the smallpox vaccine it could and may have stepped up production at home, but unless everyone is screened they will probably not ever be able to use it in the general population. Isolating the area of outbreak with focused vaccinating is the present plan. But the idea of changes in treatments, and the use of other treatments is a valid one. Most often it is the microbes' development of resistance to the drug that changes the landscape. One of the great ongoing tragedies is the development and worldwide spread of antibiotic-resistant tuberculosis, chloroquine-resistant malaria, and the HIV drug cocktail-resistant HIV viruses. They change the landscape rapidly—in time courses of years compared to development of new drugs that may take twice the time.

Of course, new targets offer the hope of drugs with no present resistance, and each new HIV drug target is celebrated. The first HIV drugs against new targets, such as Fuseon, are enjoying huge premiums.

With vaccines and antibiotics, resistance develops; new viruses appear. It affects the industry very negatively.

A drug company's path may also be diverted by the FDA. Whenever the FDA approves, say, a new diagnostic criterion for a disease, then it often means your trial will have to alter. This would have long-term benefits in the sense that the FDA is moving criteria so that it will be easier to show efficacy earlier in a trial or the measure will be more objective. But in the short term, it may mean reevaluating which patients can be in the trial and introducing new parametric measures for patients who remain in the trial. In general, however, better diagnostic criteria lead to better drug development and even to the entry of more companies into the race. Better diagnostic criteria make it clear what effect needs to be achieved and how it is to be measured. Expert workshops organized by the FDA advance drug development in important ways.

The final example, which has been mentioned before and which will be necessary to discuss further, is the situation where the FDA, by approving a competitor, may no longer allow you to conduct your trial against a placebo. The cost of this for some smaller companies may be so prohibitive that the company will cease its product development.

The landscape may also change as a result of a large population trial. The hormone replacement therapy for postmenopausal women was a standard affecting sales of not only estrogens but of osteoporosis drugs, antidepressants, and so on. When a government-sponsored study of 40,000 women is coming out and you read about it on the front page of every daily and weekly, it will affect you if you make almost any drug that is taken by postmenopausal women.

Societal changes also affect you. While the Pill would probably not be approved today, it sells well but more and more men and women want a pill for men. Viagra should not have been actively researched, but it found its indication of erectile dysfunction, common in diabetic men. Recently, some physicians have begun using Viagra and its competitors after prostate surgery with promising results.

Why Unmet Medical Needs Remain Unmet

Why do some medical needs remain unmet? It's not only in cases where the market is small—that is, not enough people have the disorder or the people who have it are poor, uninsured, or inaccessible—it's also because sometimes even a large market is not large enough for a major Pharma company.

The estimated size of the market for strokes and transient ischemic attacks is $300–$500 million, but the only drug available has to be given within three hours after the onset of the first symptoms of ischemic stroke. If you are in a big American hospital, it's advisable to have your stroke in the foyer, in order for you more readily to meet the "inclusion criteria."

Nevertheless, tPA is a major biological product by virtue of its proven beneficial effects in myocardial infarctions for which it was first approved and where it sells most. Its sales are much larger than sales of the cheaper urokinase after it entered the market for the same indication.

Everyone agrees on the high societal costs of stroke: almost always incomplete recovery, working days lost, and long-term rehabilitation. However, most Americans, even the university educated, do not recognize the symptoms of stroke, so they do not seek treatment fast enough. Without similar information and education campaigns as have been developed between 1960 and 1980 for heart attacks, society will not have much benefit even if R&D scientists succeed with a drug that works. Despite this, some large Pharma and many small ones still try even after some mega-expensive failures as was suffered by Upjohn with Trizilad that reportedly cost $1 billion.

We also would use stroke drugs in traumatic brain injury as a smaller, better defined patient group, since while they may have much larger brain damage, they are on average younger as most such injuries come from motor vehicle especially motorcycle accidents.

There will surely be new drugs in the future, but ways have to be found to evaluate efficacy by methods other than patient survival or very long-term follow-up of stroke recovery. Trial costs, especially the cost of neurological assessments and long-term follow-ups and imaging costs, are bound to restrict progress although this may change with the discovery of a truly validated good target to treat stroke. However, the scientific leads come from NMDA glutamate antagonists, caspase inhibitors, cell adhesion inhibitors, mitochondrial transition pore inhibitors, PARP inhibitors, and free-radical scavengers. Each works well in rodent models of stroke—so well that one has to sigh in despair, because the much larger human brain doesn't respond so well. Of course, in humans the drug is administered after rather than before the stroke as in most rodent experiments, and nothing works so far in the clinic.

Unfortunately, stroke will not likely be treated with a drug found accidentally. The stroke patient is not like the chronic back pain patient who has plenty of time to try every approved drug and finds that an antidepressant or antiepileptic works after two weeks of dosing. You cannot do these experiments with a stroke patient. At best we try to cool them, to maintain cranial pressure, and to block fever, which is about as much as we did for them 50 years ago. Unless they have ischemic stroke detected and diagnosed promptly enough, when they may get tPA if they come in to the hospital Emergency Department quickly enough.

The true problem is that the cost of a stroke drug for the treatment over the first 48 to 72 hours cannot exceed $20,000, and how many times can you administer it? Once, or maybe twice. It is not the billion-dollar market each Big Pharma aims at, and so stroke programs and epilepsy programs are among the first casualties of research cuts. An antidepressant may bring in five times as much and is not to be tested in a life-threatening condition. What is, however, shortsighted is that if a company develops drugs for psychiatry then a parallel program in drugs for neurology is almost free in terms of brain research expertise.

chapter

21

WHAT'S THE MOST PROFITABLE APPROACH?

CHEMICAL INNOVATION

Target-based drug discovery (TBDD) emphasizes that you want a new biological target and that a new target will enable you to treat diseases that you haven't been able to treat, or to treat them better. In this age of the Human Genome Project and other large-scale biology projects, such as the SNP consortium, genotyping the whole population of Iceland, by Decode, and Estonia, and so on, is what attracts the headlines. While this biological innovation is receiving much publicity and producing much interesting information, it is only half of what the Pharma industry does.

The Pharma industry is equally busy with chemical innovation. It takes an already clinically validated target, for which there is a drug in use, and simply tries to make a better drug. Since patients and their physicians do not care whether any improved therapeutic ratio stems from biological innovation (new target) or from chemical innovation (a new molecule to an old target), this makes a lot of sense as long as there are fewer or less severe side effects and/or a better therapeutic effect is present with the new drug. It is a better drug even if it hits the same drug target since in most diseases we do not have tolerance development as we often have in antibacterial and antiviral therapies. Therefore, using the same target with a better drug is just fine as an idea for a new medicine. In fact, doctors and patients have a smaller barrier to try it, which is why the number 2 and 3 drugs almost always outsell the number 1 drug of the same class; the trick is how to become number 3 if you miss being number 2.[136]

New "not yet clinically validated" targets carry tremendous risk, but they also hold tremendous promise because the drugs made to affect these new targets may work, for example, in currently treatment-resistant populations and so on. There is an important incidental, yet almost pragmatic, reason for working on new targets: a drug company that does not let the best of its scientists work on truly new targets will not be able to retain scientific talent. But chemical innovation is an equal portion of what every self-respecting company should do. They have to do both. This is one of the problems for the Biotech industry, which used to concentrate on biological innovation. "Professor X is an expert and cloned the gene for Y." Well that is all very nice. But what if that gene doesn't hold up as a target? The Biotech company has nothing left. Biotech was based on bioinnovation; it often had no chemists. This is changing now for small and big Biotech alike. They are all acquiring medicinal chemistry, but will they have the capital of the merged giants or at least enough capital to give them the endurance required for new small molecule drug discovery?

Efficacious in Adversity

There are not too many new targets defined by scientists in schizophrenia. A huge and costly effort has been made to bring new targets not based on animal models, which are tricky for a mental disease. How do we know when a rat hallucinates, and what does he or she think he or she is? So Roche financed and teamed up with the Icelandic genetics/genomics company DeCode, and Roche made a bundle on the NYSE listed stock, to genotype all schizophrenics and their relatives in the whole nation of Iceland—some 290 thousand people. They found that neuregulin, a quite complicated nerve growth-supporting hormone, is associated with the disease genetically. This was then confirmed on a Scottish population within 6 to 12 months! While we know this genetic linkage to be true, the Pharma industry does not celebrate because it finds neuregulin and its receptor not easily drugable. That is, it is hard to mimic a large protein hormone and get it into the brain.

[136] This dynamic might be revised in the cases of Viagra, Levitra, and Cialis.

In the absence of a clear, new drug target in such diseases, efforts have focused on improving through chemical innovation the therapeutic ratio at the existing target, in this example, the D_2 receptor. Its first well-recognized side effects were development of dyskinesia: involuntary movements mostly of the tongue and head. A true increase of therapeutic ratio required that one keep the same or better antipsychotic effect and reduce dyskinesia.[137] Further and more radical improvement of the therapeutic ratio was propelled by more careful evaluation of the true effects of the highly successful D_2 receptor antagonist antipsychotics. These indeed are robustly controlling the "positive symptoms" of schizophrenia like hallucinations and delusions, and enabled the closure of many a mental hospital in favor of community living projects. It has come to be recognized that the antipsychotics do not treat the "negative symptoms" of schizophrenia,[138] which are much more subtle and less obvious to the outsider but debilitating to the patient who is now otherwise well controlled. So risperidone and olanzapine were now marketed on being better on the negative symptoms. One had to define these, agree on their measurement, and then show in trials that the same antipsychotic effect was now accompanied by improvement in control of the negative symptoms such as emotional flattening and cognitive impairment. Companies carefully looked for a way to differentiate their new product and picked up on the budding literature on negative symptoms fast. No one speaks too loudly of the large weight gain caused by olanzapine and potential development of type 2 diabetes.

There are many established D_2 receptor antagonists. Chlorpromazine, the first antipsychotic, has many side effects, and the newer risperidone and olanzapine have fewer. All of these drugs are acting principally on the same target, though some of the side effects are through their actions at other—unintended—sites. But the chemical innovation that changed the side effect profile and created the newer drugs turned chlorpromazine into a $42 million per year drug and risperidone and olanzapine into $2 billion per year drugs.

Drug companies do not send brochures to stockholders saying: "Unfortunately, we failed to generate new revenues from biological

[137] The source of the dyskinesia was thought to involve the antimuscarinic (M1 receptor) effects of antipsychotics—thus one wanted to increase D_2 affinity over M1, and have a typical therapeutic ratio improvement over a well-identified site of side effect.

[138] The "negative" symptoms of schizophrenia are the associated low-functioning executive functions, inattention, and poor interest levels; the so-called "positive" symptoms are hallucinations and delusions.

innovation, but please find enclosed your increased dividends based on our successes in chemical innovation"—though it might be true very, very often.

The Cost of Doing Pharma Research

Whether a Pharma company is looking for a new target or trawling for chemical innovation, or, as it should be, both, it is spending some 8 to 15% of its drug development budget (R&D) on this preclinical research. The remaining 85 to 92% will be consumed by the clinical trials and the marketing. For every new chemical entity (NCE) discovered, developed, and launched, preclinical research takes about 40 biologist-years, 40 chemist-years, or at least $20–40 million at $250,000 per full-time employee (FTE).[140] This is what it costs even if you use very well-known classes of tractable drug targets like G-protein coupled receptors (GPCRs), ion channels, and enzymes. Less well-described new molecular targets may cost more in the details, but it would not drive up the overall R&D budget tremendously. A bigger barrier to the drug company could almost be classified as psychological. How much is the company prepared to invest in venturing into a new target, compared with investing in a known successful target class? If you have found, following on from the Human Genome Project, the linkage of a disease to a mutation in an important new gene product, but one that doesn't happen to be something the drug company's worked on before, then the company is very cynical about whether it will be possible to find a chemical entity—a ligand—to this kind of protein. Until that is proven and there is a viable drug candidate, it can be very hard for the innovative scientist in the business of Big Pharma. However, once a drug candidate is found either inside or outside the company on a competitor's scientific poster or patent, all of a sudden this converts from fantasy to the category of "doable, should be done by us too, and quickly."

The research, development, and marketing of a drug company are very capital intensive, but people are willing to put money into it. The

[140] No Biotech can get much cheaper a well-characterized clinical candidate with a good backup, and yet they do not have this money per project. If they sell their intellectual property earlier, there is no value added (i.e., no profit), and royalties are far down the line when sales start six to eight years later.

price to earnings ratio (P/E) returns for Pharma companies can be very high, and there is no shortage of capital. The industry is growing.

As was discussed earlier, the failure rate of drugs in development is high. Even allowing for drugs that are "entered into man" only for Proof of Principle, a failure rate of 60 to 70% of INDs is high, especially because many are just aiming to be "me toos" and are just chemical innovations. Failure rates are high for all the industry. There is no significant difference between the big companies.

The Price of Success

Even if drugs are approved and marketed, the revenues are hard won, with postapproval costs being higher than development costs. Marketing budgets in professional journals aimed at physicians have been high for many years, but the printed advertising has been extended to magazines aimed at potential patients in the general population. The parallel trend, in the United States at least, has been for more and more TV advertising direct to the consumer. Objective opinion is that this advertising often pushes the envelope of acceptability. The cautionary contraindications are clearly enunciated at great volume, but occasionally the actual indication for which the drug is intended is left a little more obscure. The advertisers encourage you to consider that you have a disease, with the underlying assumption in some cases that you may be unaware of it. Male sexual dysfunction has been introduced alongside its cure, Viagra.

Female inflammatory bowel disease was introduced with the new—in 2003—drug alosetron, with a serious contraindication being if you are of the male gender. This phenomenon is also particularly apparent in advertisements for antidepressants and sleeping aids, where stressful periods in life are being encouraged to be medicated. The unprecedented number of adolescents diagnosed with major depressive disorder (MDD) in, for example, California has its own ramifications. While adolescence in modern society is often a very difficult time psychologically, it is questionable if the number of patients in this age group should grow as fast as it does. One of the proton-pump inhibitors' advertisements has lines of people saying they feel much better, without it being clear what symptoms have been relieved. In case you have trouble typing Nexium, the web site is given as www.purplepill.com, though www.nexium.com also works. The advertisement for Propecia, which is for male-pattern baldness, gives some warnings but does not tell consumers that after taking it they will not be allowed to donate blood (at least in California).

Success May Be Short Lived

Astra-Zeneca's omeprazole (Losec) (approximately $50 per 28 20-mg tablets down from over $100 in 2001 when its turnover was some $6 billion[141]) and its esomeprazole magnesium (Nexium) ($202 per 30 20-mg tablets) are having their market positions eroded by generic omeprazole ($55 per 56 20-mg tablets) and OTC Prilosec (omeprazole) ($20 per 28 20-mg tablets).[142]

From a business point of view this is predictable. But when the largest selling drug (Losec) effectively loses its market position overnight, perturbations are felt on Wall Street. It would certainly affect badly a company that did not have a broad portfolio and whose revenues were overly dependent on one product. However, as reported in 2004 by the *Wall Street Journal*, Astra-Zeneca was regarded by analysts as having the best pipeline in that year. Within a year problems with two innovative AZ drugs Iressa and Crestor beset the company that was exalted as the best by analysts the year before. When you have drugs in large numbers of patients, anything can happen.

The consumers may not be protected by this albeit projected forthcoming drop in prices of every successful drug. The projected loss of earnings at the end of patent protection would tend to make the initial prices to consumers higher as companies try to maximize their return on research investment. The Pharma industry may be the only industry affected in this way. Car manufacturers compete on price and merits, but no one can make a "beemer" but BMW. Publishing—books, films, and music—is protected, but copyright protection is much longer and the business model competes on parameters other than being allowed to copy a product exactly.

It is, of course, worse from a business point of view when the cessation of revenues is precipitous and unforeseen as happens when drugs are withdrawn, or licenses rescinded. Some drugs are withdrawn for idiosyncratic responses, which is the most intractable intellectual problem in risk assessment. The cash-flow reverses whenever litigation occurs, *even when there is no fault of the drug company.*

The Cost of Failure

The failure of clinical candidates for a variety of reasons has already been explained. From a business perspective, the cost of failure is so great that,

---

[141] $6 billion last year prior to patent expiry according to the *Financial Times* (http://specials.ft.com/pharmaceuticals2001/FT3715MN0MC.html)

[142] 2003 prices obtained from a web search on November 2, 2003.

if you are ordained to fail, everyone wants to fail early. Hence it is very important for companies, especially if small, to try to fail early and adopt a "Smart Clinical Trials" policy. First, it is best to have an indication where you can already have an inkling in Phases I and IIa that you will have efficacy. For example, if you are making a calcitonin receptor ligand in order to treat osteoporosis, you can directly measure serum calcium in Phase I. Observing a reduction in blood-borne calcium would give you a pretty good idea that you probably won't fail on efficacy. The drug may fail on safety, but the evidence would justify your spending hard-earned dollars with some confidence.

It is possible to develop ways to ascertain in Phase I some efficacy data in endocrinological indications, antibiotics, antivirals, and certain cardiovascular indications. Even in CNS disease, for example in Alzheimer disease one can look for changes in attention span in older healthy volunteers in Phase I as an indicator of possible efficacy.

In addition, drug interactions can be studied more extensively on human microsomes, human liver cells, and so on, which makes it even more surprising that so many drugs must be withdrawn because of drug interactions: seven in the past three years. Is that lack of foresight or poor prescription policy enforcement? Are overzealous sales reps who push the good profile and neglect the side effects and liabilities believing or at least saying that "all experienced doctors will take care of that" somewhat or largely to blame?

chapter

22

PHARMACOECONOMICS FOR BIOTECH

DILEMMAS: INSURMOUNTABLE OBSTACLES OR UNACHIEVABLE GOALS OF OWN DRUG DEVELOPMENT, AND THE DREAM AND REALITY OF IPO

What does it take for a Pharma or Biotech company to remain viable? To be fiscally self-sufficient? What would a small, middle-ranking drug company generating some $5 billion per year—that would place it at around #15–20 in the list of drug companies ranked by turnover—need to produce to maintain this revenue, without growing? It would need an average of one to two significant new chemical entities (NCE) launches per year. To fuel this output, with an expectation of 10 to 20% success, its R&D would need to deliver 15 to 20 clinical candidates per year. And for this output, it has to complete 40 to 80 preclinical research projects aimed at lead optimization per year from a pipeline of some 300 drug discovery projects, each of which is on a four-year cycle. Some of these projects will be new, but many will just be striving to produce new drugs for old, familiar, validated targets. This is just to keep afloat, not to grow. The difficulty is that many, many projects lead nowhere, with 90% of projects failing in the clinic. For the Mercks and Pfizers of the world, you have to multiply these numbers by almost three to five.

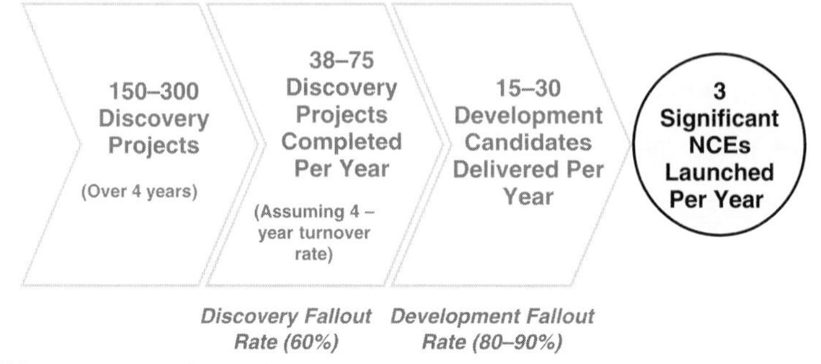

150–300 targets are being worked on simultaneously; several are old ones (using chemical innovation) but most are new. This is the requirement with the current attrition rate

Figure 22.1 Number of Discovery Projects and Drug Targets Required for a Medium Pharma Company with 10% Growth Target

From a business point of view, it is hard to manage programs with an overall 1% success rate. When Merck announced (in 2003) it was not going to launch a new antidepressant—aimed, incidentally, at the novel target, the NK-1-type substance-P receptor—Wall Street reacted severely. Although this was not the withdrawal of an approved money-making drug, it was one believed to be firmly in the pipeline and scheduled for launch, and to make $1 to 2 billion/year three years from then. So it was loss of a dream not of real revenue, yet the valuation of the company fell, as we mentioned before, by about $5 billion, and new reservations about the strength of the entire Merck pipeline surfaced.

Share Risk: Partner with Big Pharma

A strategy for new Biotechs and smaller Pharma is to spread the risks and financial burdens with Big Pharma:

- Share the cost and risk of clinical trials—and, of course, lose most of the potential upside—making sure that they license the product of their preclinical research to a company very experienced in the therapeutic area or clinical field.
- Work with another more experienced company during the approval process.
- License its product for manufacturing—the manufacturing of sufficient quantities for large human trials is demanding.

- License its product for sales and marketing—particularly for marketing in another international territory.
- Develop other revenue sources such as through acting as a "platform company" providing specific and proprietary services to the Pharma industry in diagnostic and testing paradigms.

Irrefutable Statistics: Not Every Drug for an Indication Will Make It

Since it is well known that not every drug for an indication will make it, why does Wall Street react so swiftly and badly to setbacks? In the specific case of Merck and antidepressants, it is for two reasons, the first of which is indirect.

The first reason is that while the market for an indication may be seemingly saturated, there has been historical precedence for success in a saturated market; thus, Wall Street's expectations are built upon optimism. For any given indication the first three NCEs in a class have traditionally done well. How many D_2 antagonists can the market bear? For schizophrenia—a reasonably well-described indication that was fully diagnosed and largely treated in the United States and the West—it was not predictable that there would be an opportunity to create a new $1 billion per year drug acting on the target, but both Johnson & Johnson and Eli Lilly did it with risperidone and olanzapine, respectively, within two years of each other.

An optimistic extrapolation to this accompanied Merck's foray into depression. Merck had created a lot of news by recruiting over the last decade of the twentieth century very prominent neuroscientists to head up new R&D programs for neurological and psychiatric disorders. The expectation was that results would follow. Entering a saturated market would have been regarded as a positive move. The so-called Monaminergic Theory of Antidepressant Action had, through three drug discovery paradigms, given birth to a $15.2 billion indication. Surely there was room for one more? Conversely, how could Merck afford not to be in the market of drugs for major depression, a very large indication?

Since new molecular biological and transgenic approaches have revealed a new group of targets for antidepressants, the substance-P receptor NK-1 (see Box 22.1 for details), antidepressants are still seen as a good business investment.

Box 22.1 The Birth of the Largest Indication: Major Depression: $15.2 Billion

Three drug discovery paradigms pointed to monoaminergic targets and to the NK-1 receptor to treat major depression

- Paradigm 1a: Side effect of known drug
  - Iproniazid: the putative anti-TB drug: improves melancholy but does not improve TB; it acts as an inhibitor of MAO A/B that metabolizes NE, DA, & 5-HT
- Paradigm 1b: Natural product
  - Reserpine precipitates depression-like syndrome—empties NE & 5-HT stores (implicates monoamine deficit)
- Paradigm 1c: Pathophysiology
  - Extremely low 5-HIAA in suicide victims' brains (Asberg) suggests 5-HT deficit
- Paradigm 2: Molecular and cellular models
  - Tricyclics & SSRIs inhibit NE & 5-HT uptake in cellular model (platelets and synaptosomes), the transporter was cloned long after the clinical trials with 5-HT uptake blockers (1973–1991). Zimelidine, Fluoxetine, and the "pack": 6 SSRIs
- Paradigm 3a: SP mRNA and gene product levels are elevated in depression models
  - NK-1 nonpeptide antagonists in the clinic—from five companies
  - NK-1 receptor is expressed on NE & 5-HT neurons
- Paradigm 3b: No new targets from genomics and proteomics studies yet
  - Transcriptional profiling on autopsy material from suicide victims and from chronically depressed patients has suggested some new targets

How did the antidepressant drugs emerge into the market and create a successful market? Scientists didn't have high-throughput screening to test hypotheses. Drugs came from good clinical observation. After the war people in clinics for tuberculosis (TB) were treated using iproniazid. It didn't cure the TB, but their melancholy improved. Since iproniazid was a known inhibitor of the enzyme monoamine oxidase A & B (MAO-A, MAO-B), which metabolizes the monoamines—norepinephrine (NE), dopamine (DA), and serotonin (5-HT)—acting as neurotransmitters, the monoamine hypothesis of depression was born. The other way of discovering the link between monoamines and depression would come through finding a natural product. An expedition to the Amazon basin brought home *rauwolfia* or reserpine. It is an alkaloid from a plant that precipitates a depression-like syndrome by emptying norepinephrine stores in the brain. So, a specific monoamine deficit causes depressive syndromes, and iproniazid's antidepressant effect comes from increasing monoamine concentration. These are compatible observations. Controversy comes from arguments about which monoamine is the most important. A parallel finding comes from pathophysiology: Marie Asberg looked at 78 people who had committed very violent suicide, for example, by jumping from a bridge, or throwing themselves in front of a train. These profoundly affected patients had extremely low levels of the metabolite 5-hydroxy-indole-acetic acid (5-HIAA), which is a metabolite of serotonin (5-HT). So it suggested that serotonin itself was low in very depressed people.

Box 22.1 The Birth of the Largest Indication: Major Depression: $15.2 Billion—*cont'd*

Inspired by these observations, and based on cellular and molecular models, everybody was making analogues of chlorpromazine to make a better antipsychotic; they accidentally made a tricyclic, which turned out, in the same psychiatric clinics, to be an antidepressant. All the tricyclics, amitriptyline, and so on, came from the chlorpromazine development programs.

The third wave of antidepressant research comes from more modern thinking. People looked at molecular signals at the genetic level—such as messenger RNA (mRNA) activity using microarrays and gene product levels—and it turned out that the substance-P receptor, which is expressed in noradrenergic neurons, is somewhat upregulated, but more importantly, there are big increases in anxiety and in depression models in the substance-P level. So substance-P antagonists to the subtype of the substance-P receptor NK-1 are now in clinical trials with five companies because they want to join the six companies and drugs that currently share the $15.2 billion market.

Also looking at the business of mental disease objectively, but without cynicism, a common denominator of these indications is that they share the distinction of not being cured by these pharmacological treatments. This makes the market even more attractive. The patients have to take the drugs chronically. Not only are the diseases not cured, but there are few treatments that give 100% relief to those who have a syndrome. All usual response rates are 60 to 70% for a really good drug. For example, SSRIs are very good antidepressants, but the response rate is 60 to 70% even if you accept all the side effects of, for example, sexual dysfunction. This gives a double opportunity: (1) one can enter a partially saturated market with a drug that works on patients unresponsive to existing treatments; and (2) one can improve on the side effect profile or the efficacy in terms of the time required for the onset, which is 14 to 20 days for SSRIs.

Antipsychotics are in this way very similar to antidepressants. Chlorpromazine, haloperidol, and all the rest, are fantastic breakthroughs, but there are 30% of psychotic patients who do not respond to any, and we cannot control their psychotic behavior. A new mechanism of action may help with them. Again, while all antiepileptics are highly effective drugs in responding individuals, there are 30% of patients for whom we cannot provide anything to control their seizures. There is plenty for pharmaceutical companies to do, but the risks are higher with new targets and new mechanisms.

Science Fiction to Fact

Another way to survive with fewer resources is to accelerate the drug development cycle. It normally takes approximately 10 years to develop drugs. How can this be done in four years? A three-year clinical trial cannot be done faster than three years, surely? And "a pregnancy also takes nine months for those in hurry," to quote a clinical development head of a prominent Big Pharma company. But, faster than possible can and has been done. For certain biologicals, therapeutic antibodies with targets provided from the known biology of inflammatory cytokines—TNF, IL-1, or other antigens—come from proteomic analysis of a tumor type and have had development cycles of about four years. The work from the laboratory is taken into humans quickly by making human antibodies or humanized antibodies, which are themselves the drug or can be made to carry a drug to a specific target. These are safe. They are often efficacious.[143] But they are difficult to administer; you have to inject them. More importantly from the business perspective, they mostly work for small, fragmented indications—usually specific cancers—and, thus, they are perfect for small companies. The first Biotech company in San Diego to make money from product sales, IDEC Pharmaceuticals, did just this with therapeutic antibodies.

Target Feast and Famine

Cancer research provides many targets and leads to many clinical development candidates. Cancer is split into many small distinct indications, all of which require distinct therapeutic strategies. There are many, many targets specific to individual cancers, and these targets are being revealed by genomic and proteomics research. Despite all this knowledge, the number of oncology drugs approved in 2002 was roughly the same as that approved in 2001: four! Out of those, three were biologicals (i.e., antibodies), and one was Gleevec, from a very large company, Novartis, which, for good measure and to be on the safe side, also had a therapeutic antibody approved the same year: Solair, which acts against IgEs formed in allergy. But this number was really expected to grow in 2003 and 2004 *et seq.*, even though it was very difficult to go from target to medicine.

At the other end of the spectrum, for psychiatric disorders such as schizophrenia, research provides few targets for candidates. This

---

[143] For example, as was seen with Herceptin (see Chapter 17). Several hundred Biotech and Big Pharma developed therapeutic antibodies in oncology where trials are fast and improvements of expected life length by just 6 months are celebrated on the first page of the *New York Times*.

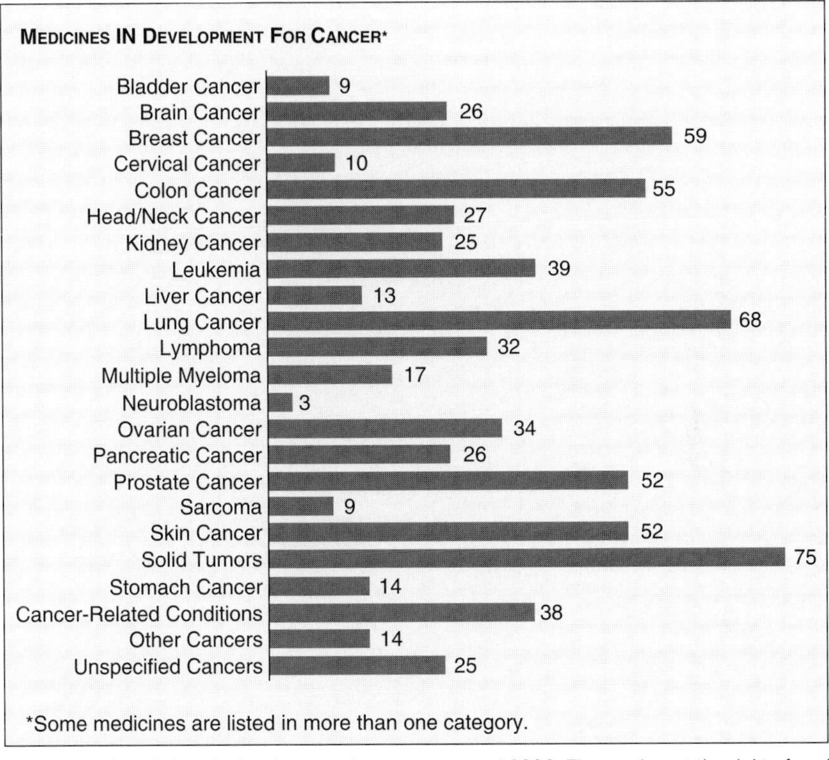

A snapshot of medicines in development for cancer around 2002. The number at the right of each bar is the number of targets pursued at the time. There is overlap between the categories, but of these identified 722 targets we may only see four to five drugs emerge.

Figure 22.2 Medicines in Development for Cancer

is despite the fact that schizophrenia is a 20 times bigger indication (i.e., 20 times more people have it) than any one of the individual cancer forms. This, of course, does not make it 20 times bigger economically, despite the average schizophrenia patient taking the drug for 35 years daily, while no cancer patients have this opportunity or luck with their drugs. For schizophrenia, there are only 16 trials ongoing and 14 of them are dopamine antagonists and the 2 others are nicotinic α-7 receptor antagonists, based on the anecdotal evidence that *schizophrenics smoke more than the general population and, perhaps, that represents a form of "self-medication,"* as well as proven effects of nicotine on attention and other "negative symptoms" of schizophoenia. Now how good is this for the validation of a target? The animal models to back the nicotinic α-7 agonist are nonspecific behavioral models in rodents,[144] since we have

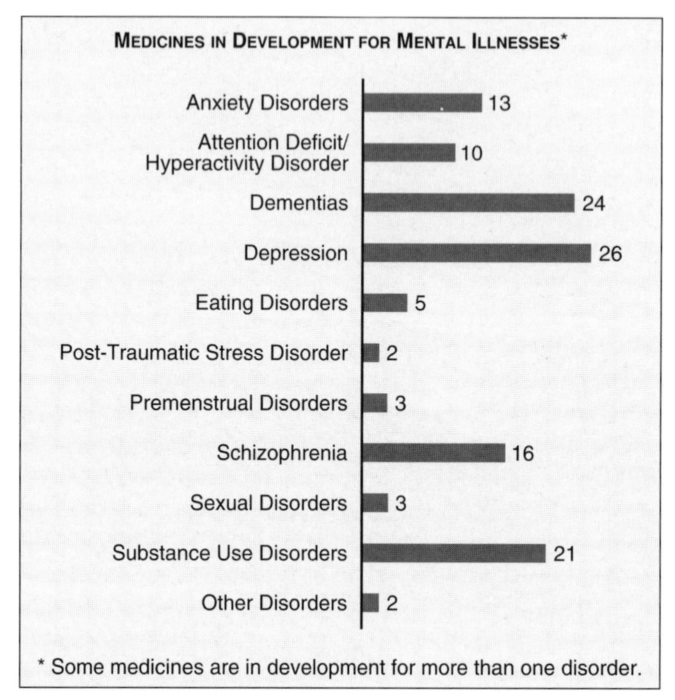

The 125 clinical trials in process for mental disorders (c.2002).

Figure 22.3 Medicines in Development for Mental Illnesses

no idea what a psychotic rat is, how he or she feels, and about what he or she hallucinates. The situation with cognitive deficits is easier to measure. The nicotinic α-7 drugs have clear effects on cognition in rodents and in man, which is known from a stopped trial.

THE SCIENTIFIC BASIS OF DRUG DISCOVERY IS EXPANDING

From a business point of view, there is good reason for cautious optimism that the drug industry can continue to grow as the number of targets grows thanks to genomics and to better understanding of pathways. The number

[144] The nonspecific behavioral model paradigm is by "prepulse inhibition" that detects many antipsychotic drugs but also other drugs.

of clinical candidates will grow. Most business experts are astounded by the sheer number of potential molecules with therapeutic activity. There are many hundreds of drugs still to discover, in particular if medicine becomes more personalized. Think that there are over 1,200 jeans models to suit the lower part of your body for a few months or year, depending on fashion, but there are less than 100 drugs to treat your heart, which works all the time for 85 or more years.

Future drug targets are expected to be discovered. For the moment one can safely predict that by 2010 the biological Proof of Principle will exist for around 600 targets and the number of new chemical entities (NCEs) will be 1,200 leading to 5,000 to 15,000 drugs. Not all new drugs will be that different from each other but all of this will happen if there are no major legislative or economical events that change the present trends.

Are Biotechs Smarter?

This is perhaps a more important question for Biotech to ponder than the business community itself. We have already noted that Biotech needs to have a higher success rate than Big Pharma; why ask the question of their being smarter?

A company like GlaxoSmithKline, built on the mergers of Glaxo with Wellcome and then with SmithKline Beecham, with research labs in the United States and United Kingdom was in May 2002 conducting 24 Phase I and 14 (4 oral, 10 parental) Phase II clinical trials, and another 10 vaccine trials (4 Phase I and 6 Phase II). This is 38 to 48 trials for its workforce of some 14,000 scientists. Biotech companies try to do one to two trials with an average of 70 scientists. This would make them about five times as efficient. Big Pharma is also conducting research into 300 to 400 compounds to launch just one, two, or, hopefully, three drugs per year; then a Biotech with just one or two clinical candidates needs to be 150 to 200 times as effective. Since the scientists in both types of companies have all had equivalent training, it is hard to believe that they will be even 5 times as efficient and 100 times as effective. They may be twice as smart—maybe. All Biotechs are proud to present the few guys they recruited from Big Pharma, intentionally and calculatingly implying to the investors that they will work like Big Pharma. If they succeed in doing it with much smaller resources, that is already admirable, but can they do it 10 times better with one-third of the resources per project just because they have a higher chance to hit the jackpot if they succeed? Sometimes, but not reliably so. Neither Big Pharma, which reduces risk with an increasing number of projects, nor Biotech are even safe bets in terms of productivity. As mentioned according to Fortune magazine, Biotech investors so for have lost $45 billion. However, those who bet more luckily own stock in some of the 157 biotechs still alive and quoted on Nasdaq.

Even if one takes into account that with biologicals the drug development time may be reduced, Biotech's competitive edge may be short-lived as Big Pharma recognizes the future of biologicals and moves into Biotech's arena, as, indeed, it has done so well either by acquisition (Genentech is controlled by Roche), by co-marketing (Gilead-Roche-BMS), or by simply competitively beating Biotech in biologicals. It is becoming easier as the main human antibody companies Medarex and Abgenix would rather work with an antigen from Big Pharma that can pay for the work up front, and that will be more able to do the trials and market the drugs than would Small Biotech. Moreover, these two companies sit on the key patents.

Biotechs are largely amateurs when it comes to successful product development, which may be the reason they have had so many financial woes. When Biotechs successfully raise big money through IPO, that is when they truly realize that their big money is small money because, to belong in the league of Big Pharma, you need to spend about $2 billion or more a year on R&D. The most they raised was close to this sum. The exceptions to the rule, Genentech, Chiron, and Amgen, are still struggling over whether to have or not to have a small molecule drug discovery program, and they start and stop these because they do not have the reserve, experience, and pockets to stay the course—until they feel "rich" again and restart a chemistry-based program, that is. All three did this. Newer-comers like DeCode and Celera, which raised a lot of cash to try to play at being Big Pharma, have bought their own medicinal chemistry. Their management is increasingly recruited from Big Pharma. The executives vow to do "it" more efficiently, but how many times more efficiently can they perform while using the same brain and same databases? Are they chanceless and beyond risk? Of course, it is not that there is not plenty of room to make new profitable drugs, but more that it will be hard to become a global company with its own development and marketing without $5 billion or more in sales. We know the tale of those who asked Pfizer to co-market with them. Warner-Lambert with Lipitor was bought up later; as was Pharmacia-UpJohn-Searle with Celebrex. Is this bad? Not necessarily, but there is a fairy-tale quality to it. If it is a big enough product, you cannot pay the royalties on it. The exception that proves the rule is exemplified by Astra buying back the U.S. co-marketing rights to Losec from Merck for billions of dollars, a deal that is generally considered as a blunder by Merck. Now Astra-Zeneca is larger than Merck.

chapter

23

SHRINKING VALUE OF TARGETS

INFLATING VALUATIONS

Almost in line with the enthusiasm that greeted the Internet business revolution with the inflated value given to business initiatives without a real prospect of revenues or profits, many entrepreneurial scientists put an extremely high price on their targets and candidates, and either set up Biotechs galore to exploit their dreams or expected Big Pharma to buy the rights to their targets, and in the few cases where they had the chemistry, the rights to their compounds, at amazingly high prices.

While, for the most part, Big Pharma didn't get involved in bidding wars for compounds, this was by no means universal. Perhaps many Biotechs benefited from the largesse of Big Pharma investing in their ideas. There are real costs incurred by paying for compounds, buying companies, or contracting companies to find targets and molecules, which can have a negative effect on the bottom line.

The very real problem with buying any idea or single chemical entity is that it is wholly likely not to succeed, and paying up front much more than one would pay for a clinical candidate developed in house is unsound business practice. Of course, since a molecule can generate billions of dollars, it can be very tempting. What is the real worth of targets

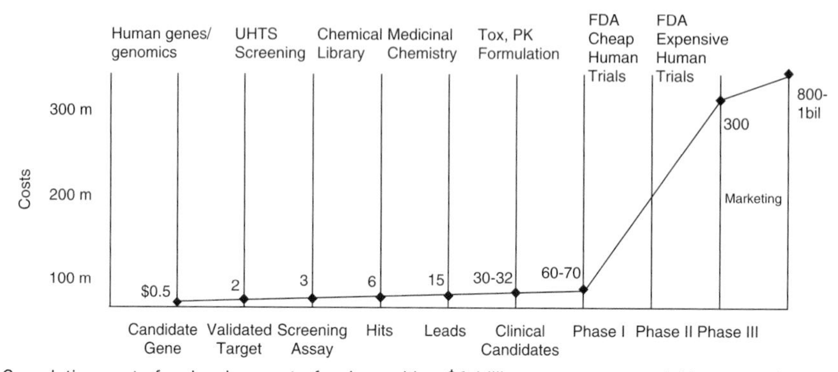

Cumulative costs for development of a drug with a $1 billion per year potential increase dramatically after Phase I. Only 10% of these drugs that entered into man "survive."

Figure 23.1 Costs for Development of a Drug Using a Target-Based Drug Discovery Paradigm for a Large Clinical Indication

and clinical candidates? To ascertain this, one looks at costs and, at the same time, projected revenues.

The cumulative costs for drug development are shown in Figure 23.1. The cumulative costs after the molecule has become a clinical candidate become exorbitant for a small company. If a company then has to market the drug, one has an almost limitless opportunity to increase expenditure. Should a small Biotech take its drug through to Phase I and accept the risk of failing? If it wants to realize the value of its research, then it must. The first real value inflection point occurs when you have a compound that has been in man: it is safe; there is some dosing data about it; and, in lucky cases, also some indication of possible efficacy. But Biotech—like the big companies—should only do so if it has more than one Phase I compound. There should also be backup compounds to its leading candidate, as well as some compounds for another target to balance the target and the compound-related risks. If you are a small Biotech, it is immediately much better to become a large Biotech, which makes it more surprising that some scientists always want to form their own company rather than develop their ideas by entering into an equitable arrangement with an existing Biotech.

If a Biotech company does decide to sell its drug rather than develop it itself, what is the sensible expectation for a drug that will generate $1 billion per year? Here the calculation is based on the accumulated value of the drug as it is developed. If you are a small company and you have only a single target, then you can get $100,000 for it and expect very little royalty—say 1%. Your cost, on the other hand, is about \$5 million. If you bring it all the way to a clinical candidate, that is, you have developed a suitable chemical entity that works in animal models *in vivo*, is not toxic, and has good pharmacokinetic properties. These are big musts, hard to achieve, and you need to score on all of them. You can then get \$3 million and a 3% royalty. Your cost is now about \$15–20 million. When you can deliver a compound already in Phase I or even Phase IIA then you FINALLY can get more for the compound than you have actually spent developing it. Your costs have ballooned to about \$20–30 million. But to get this far, you needed investors who understood this and gave you \$30 million per such compound, and a minimum of \$60 million over the years for you to avoid history recording you as a single-drug company.

If you only have a target to sell, these have become very cheap because of the Human Genome Project. Given the new and publicly available data, the worth of an individual nonvalidated target is becoming close to zero. A validated target in a relevant animal model may still have some value, about \$250,000, depending on the patents and data associated with the target. Ten years ago you could sell targets or platform technologies to discover targets; you cannot anymore in 2005.

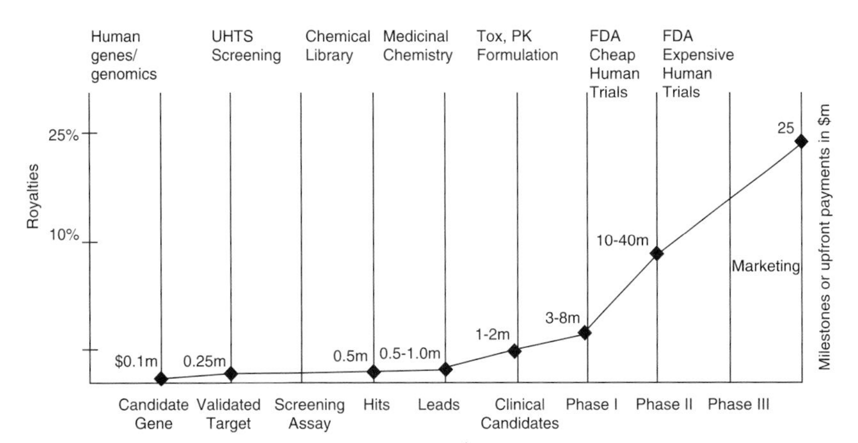

Up-front milestone payments, and royalties for a \$1 billion indication in, say, cardiovascular or CNS. The accumulated value of a drug increases according to where it is on the development path.

Figure 23.2 Royalties and Milestone Payments for a Large Drug with Expected Sales \$1 Billion/Year

New Business of Wiser Biotechs

If you are, or want to be, a growing Biotech, then having your own medicinal chemistry is the best medicine for your targets. The new kinds of Biotech companies are formerly solely genomics-based companies that are looking to acquire medicinal chemistry. These include Millennium, DeCode, Genset, and Celera/Lion Biosciences. Similarly, from the other direction, combinatorial chemistry companies, such as Pharmacopea, are trying to embrace biology and become Biotechs.

The emerging companies that result are going to be much better placed to find targets, discover clinical candidates, and take them through to Phase I and Phase IIA, and get a much better return on their intellectual and financial investment. The new Biotechs will also need longer time and more investor money to do this. So venture capitalists will shave their portfolios to select fewer companies and push them longer, thereby inevitably increasing their risk doubly. To make matters worse for Biotech investors, IPOs in 2004–5 did not really raise enough funds to make manyfold returns on initial investments made in the few Biotechs that made it to IPO (by having one to two compounds or biologicals in the clinic and having spent $50–$100 million).

There is also great hope that these companies will make new—truly new—drugs.

chapter

24

ASSESSING COMPANY ASSETS? LOOK IN THE LIBRARY

HIGH-THROUGHPUT-SCREENING-BASED DRUG DISCOVERY PROCESS AT BIG PHARMA

An appreciation of the assets of a company comes from understanding the processes involved in drug discovery. These have been outlined earlier, but some historical tribute to the origins of a company's major assets merit being emphasized and appreciated, even if one is principally interested in the business as a financial tool.

High-throughput screening is the modern way of testing many, many compounds against putative targets. It is getting better and better, and even better. "Better" means that you have more compounds, and you need even smaller quantities of these compounds in smaller and smaller volumes, and the assays become optically readable, meaning that they are so fast you can read a million data points a day. We can now test 330,000 compounds in triplicates against a target protein in a single day. Therefore, you can do a lot. This means that if you have a good validated target, either identified clinically or in a good animal model, of a drugable class, you do not have to think too carefully before trying to find a suitable candidate drug to affect the target. This is a great achievement dependent on a major advance in chemistry. It stems from the conception and creation of combinatorial chemical libraries by chemists such as the Hungarian Árpád Furka and the German I. Ugi. Chemists such as the Nobel Laureate Barry Sharpless introduced "click chemistry" to assist in the making of huge numbers of efficiently synthesized compounds to make large libraries. These large libraries of compounds now have to be tested against the target protein. Thus, one needs the means to produce large quantities of the target protein, preferably by growing cells that express the target protein. Finally one needs micro- and nano-fluidics robotization to mix small volumes (i.e., 1 millionth of a liter or smaller) of cell products and the compound, and then one needs informatics technology to collect and store the data. The many companies—such as Evotec, TAP, Calypsus, and Siemens—that developed integrated drug and target screening systems have had good customers in Pharma.

The heart of the drug discovery process is the screening of chemical libraries using especially developed assays. The asset of a company can be measured in its breadth and depth of its chemical library and its ability to use the library with carefully constructed assays to identify hits against prospective targets.

How do companies identify a target that they can truly believe in? Genomics and proteomics may identify a prodigious number of potential targets, but how do you know that if you address or affect such a target you will have a therapeutic effect? The validation of targets is a key responsibility of the biologists. They can do it by finding clinical data from pathophysiological changes; they can find it from genetics; they can find it from animal models using genetic tools, such as transgenic animals, antisense, or siRNAs; but best is when they can validate the target with a chemical which is not perfect yet, but selective enough that its effects can be regarded to be mediated through the target, and only through the target you want to validate. With such compounds you hit two birds with one stone. You can show that the occupancy or inhibition of your target produces the desired effect, and you also prove that not only is this protein a good target but molecules can be found to hit it selectively. This must be possible since you already have one! Many times chemists will not give you the benefit of the doubt that your target protein is drugable, but a single small synthetic molecule changes their mind 100% of the time.

The dream for a chemist is rational drug design where the three dimensional structure of the protein is known in detail to angstroms or nanometers, and now you can dock existing and imaginary compounds by the millions into the binding site(s) of this protein, first on your computer—if you have the computing power and the software at only a few millions of dollars—and then in reality by making or simply selecting from your existing library the best fitting compounds. How do you make a large and diverse library with a very large number of chemical entities that are likely to be meaningful ligands to proteins of interest? Companies have a few hundred thousand compounds in their historical libraries as a legacy of earlier projects. Making 100,000 compounds up to the late 1960s would have been considered a lunacy, as they were then tested in whole animal assays, or in cellular and organ and tissue assays, with very low throughput. To create the millions of compounds chemists have used in the past 12 to 15 years, one needed modern methods of combinatorial and parallel chemistry as envisaged by Ugi, Furka, and the others?[145]

Finally, how does a company select a clinical candidate that really goes to the clinic and on which they can start to spend serious money? Ultimately, the demonstrated ability to go from idea to clinic is how you judge companies.

When it gets down to "screening" for potential hits, the library determines your success. The chemical diversity and the quality of the library are what will ultimately determine the potential depth and breadth of the company in question. Combinatorial libraries have had their share of praise and criticism. Their value will be judged when we identify which libraries produced the most valuable screening hits over a 10-year period. Today's tally shows that historical and natural product libraries are standing as better sources of hits to traditional targets than combinatorial libraries of the late nineties. "Pharmacophore" libraries and other variants of new combinatorial libraries are also starting to contribute to screening hits.

Hits against Receptors

How might we compare companies at the molecular level? Data are not always easy to come by. The data in Figure 24.1 illustrate an important point. The question was how many compounds—if any—in the libraries of which companies will bind with one micromolar or higher affinity with these receptors? Incidentally, you do not have to know the endogenous ligand to your target before running it against the library. One can run as a target an unidentified orphan receptor. After finding hits, one can go back and identify the receptor once interest has been elevated by finding hits

[145] For review on fathers of the field, see Michal Lebl's Parallel Personal Comments on "Classical" Papers in Combinatorial Chemistry, *J. Comb. Chem.*

Receptor	Size of peptide (aa)	% of screened compounds 1 μM or higher affinity	
NK1	11aa	0.5–1	4 companies
Galanin R1	29aa	<0.001	8 companies
CRF1	41aa	0.1–0.5	4 companies
Orphan	19aa Nociceptin	1	2 companies

The frequency of hits when four example receptors were run against the proprietary libraries of up to five companies for each receptor. The specific molecules that combined with the receptors—or targets—are not given. Each of the companies had hits against the NK-1 receptor implicated in depression; no company had hits for galanin (save for low affinity hits from Johnson and Johnson and Schering); some companies had hits against corticotrophin releasing factor (CRF). In the last example two companies ran an unknown "orphan" target against their libraries. Having obtained hits, the companies later identified the target as nociceptin, an important receptor for pain.

Figure 24.1 Frequency of Hits in Random Screening for Peptide Receptors and for Orphan Receptors in Historical Pharma Libraries

against it. This is behind the technology of some Biotechs like Arena Pharmaceuticals and Acadia. One uses the new artificial ligand to explore the biological and pharmacological relevance of genomics-derived targets, and to provide a chemical Proof of Principle for a new target.

WHICH TARGET SHOULD ATTRACT INVESTMENT?

What should govern corporate strategies in a fragmented market? The market is by definition fragmented because no company has more than 8 to 10% of the total market. For a therapeutic area such as cardiovascular disease or oncology or psychiatry, however, there are companies that have up to 25% of the market.

Drug company decisions about molecular targets are made at several levels. One needs to start with the selection of the indication by objective assessment of the therapeutic areas, and within the therapeutic area, which indications can be treated. Target selection takes into account whether one is aiming to treat the symptoms or to affect disease progression or prevention, or, regrettably unlikely, a cure. The companies assess the market opportunity at the exploratory, advanced, and clinical project levels. In other words: does the prevailing scientific thinking present opportunities; can these be developed; and does your company have the ability to take these ideas into the clinic?

If there is a symptomatic corporate malaise in the industry, it might be that companies are more attracted to areas of success for their competition. A consequence of this is that only 2 of the big 20 companies are in dermatology, only 3 in antibiotics, and only 2 making vaccines. Naturally, neither of these therapeutic areas—dermatology nor antibiotics—is as large a money-maker as cardiovascular or CNS drugs, but Allergan ekes out a living in dermatology and ophthalmology. This means that some therapeutic areas are very neglected and provide opportunities to smaller companies. Oncology is a fragmented area where Big Pharma works on general pathway-based drugs that may work in many cancers, for example, Iressa (gefitinib) from Astra-Zeneca, an epidermal growth factor receptor (EGFR) inhibitor for solid tumors, or in large cancer forms such as prostate or breast, whereas Biotech works on smaller more rare cancers preferably and uses biologicals most often. There are of course exceptions to this simplified description. The 2001-launched Gleevec (imatinib mesylate)—the first tyrosine kinase inhibitor to become a drug—was for a small indication (chronic myeloid leukemia,

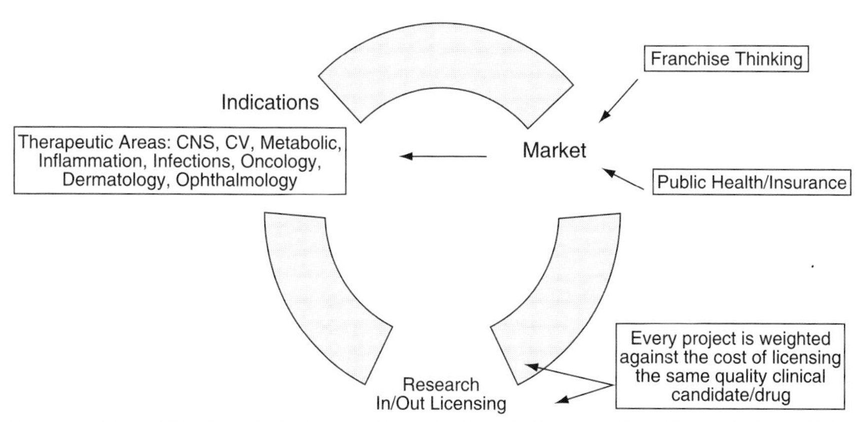

The selection of the target starts with the selection of the indication. The selection of the indication comes from competitive market assessment, therapeutic areas with indications, and assessment of research and development opportunities both in house and out-of-house.

Figure 24.2 How to Select a Target

but gastrointestinal stromal tumors (GISTs) may be added as an indication) and it came from a Large Pharma, Novartis.

Franchise thinking is an important and constantly debated part of Pharma decision making. If you have a sales force that knows every cardiologist in the country, costs more than your total research organization, and brings in about 25% of your revenue, can you afford not to have a drug in every possible new area of cardiovascular medicine? No. If a small company now discovers a new drug in this area, you want to be the one they call first to propose collaboration or joint venture. Because you have the best clinical trial record, and because you have the best marketing muscle, the royalty to the discovering company will be worth more. But does this mean that as a newcomer in the field, you are chanceless? No! As discussed earlier when Astra, with help from Merck in the United States, launched its revolutionary Losec (omeprazole), it had no drug to treat ulcers. The other two big ulcer drugs were each making a billion dollars per year for Glaxo-Wellcome (Zantac) and SmithKline Beecham (Tagamet)[146] and their market positions were vigorously defended. Yet Losec became the largest selling drug, almost fully eliminating gastric surgery for ulcers, thus dramatically changing medical practices. So franchise is nice to have, and worth defending but is not a must in this business because doctors and patients will go for the best. Faithfulness or loyalty to a label or brand is not as prevalent as in the car industry.

[146] When Glaxo-Wellcome and SKB merged, they were allowed to keep both antacid H_2 blockers because of small market share at that time.

chapter

25

To Merge or Not to Merge?

What Drives Mergers (Part I)?

The Pharma industry emerged over 120 years ago as a result of dye and chemical companies' efforts for a more sophisticated product line. Making drugs instead of dyes to color uniforms gives extraordinary added value. Some of the original companies such as BASF are still going; other names—such as Ciba—have been lost in a series of mergers and acquisitions. Other names have been brought forward again as the parent company's reputation was maybe somewhat tarnished, as happened when American Home Products changed its name back to Wyeth following the Fen-Phen[147] debacle. Many of the industry's names have disappeared. Despite some spectacular setbacks with individual medicines, the overall medicines market has increased continuously in the past 50 years by doubling every six years. There is

[147] Fenfluramine (approved in 1973, withdrawn in 1997) and phentermine (appetite suppressant approved in 1959 and still available). Wyeth-Ayerst Laboratories, a subsidiary of American Home Products Corp. of Madison, New Jersey, manufactured and marketed fenfluramine under the brand name Pondimin. Wyeth-Ayerst also marketed Redux (dexfenfluramine), which was manufactured for Interneuron Pharmaceuticals. See http://www.fda.gov/cder/news/phen/fenphenpr81597.htm

a distinct willingness to invest in the capital-intensive Pharma industry. Objectively speaking, Pharma is not a mature or consolidated industry. It has many players; none has more than 5 to 7% of market share. Thus, mergers are logical. A closer look shows that if one looks at therapeutic areas, then the picture is no longer so fragmented. The truth lies in between.

While some mergers have undeniably been rewarding to ambitious executives of the companies involved, the main business reason for merging is a response not only to the need for companies to grow but also to the continually increasing costs of drug discovery and companies' needs to distribute the risk and amass the capital required for research and development.

Growth and mergers and acquisitions have progressed in parallel over the last 15 years with, for example, Roche-Syntex, followed by Glaxo merging with Wellcome, then with SmithKline Beecham, Sandoz-CibaGeigy (Novartis), Bristol-Myers Squibb DuPont, Rhone-Poulenc-Rorer and Hoechst (Aventis) now with Sanofi-Synthélabo, Astra-Zeneca, etc. Pfizer is the most acquisitive having absorbed Parke-Davis, Warner Lambert, and now Pharmacia-Upjohn, which had itself earlier absorbed Donald Rumsfeld's old company, Searle. A lot of the acquisitions are tactical in order to gain access to a market or a particular drug—for example, Pharmacia buying Searle for its Cox-2 inhibitor Celebrex (celecoxib)—or to cement market position. Pfizer probably bought Parke-Davis/Warner-Lambert (WL) for Lipitor because it needed to consolidate its cardiovascular market position with a cholesterol-lowering drug. Pfizer also stopped most WL R&D projects, or so is it recounted. The executives and R&D teams are in constant flux. Merck, on the other hand, has been largely relying on organic growth—something Pfizer would consider too risky—and this approach deserves objective assessment in

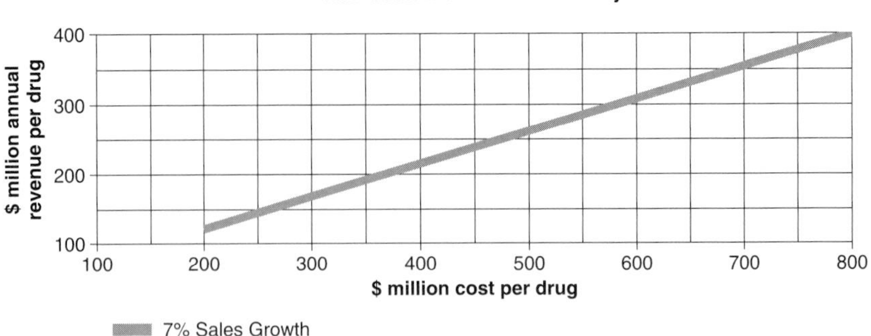

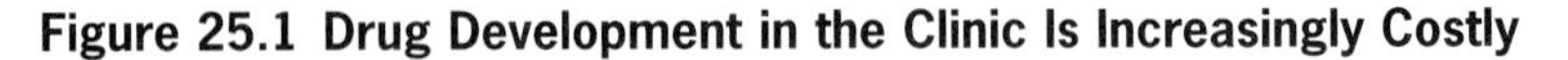

Figure 25.1 Drug Development in the Clinic Is Increasingly Costly

terms of the industry as well as the company. Merck's policy flies in the face of the anguished cries of analysts, who can now legitimately mourn Merck's bread-winning statin Zocor "going generic." As we said earlier, it's not worth remembering the names that result from all the mergers because tomorrow it will change.

The real evaluation of whether these mergers were good or bad has not been done yet. It is clear that these evaluations must start with the corporate economy since that is the driving force—not science, patient groups, or regulatory agencies. However, contemplation of and concern for potentially monopolistic takeovers plays an important role in companies' consideration of what drugs to keep and which to spin off after mergers.

There are, of course, recognized downsides. If you acquire a company for a particular product in order to enter a market or indication, you will also be acquiring a corporate knowledge and R&D team. However, part of the rationale to merge comes from cost savings; therefore most departments will be contracted—downsized—and many of the developments of the acquired company will be dropped in favor of the stronger parent companies' own plans. Corporate knowledge will be lost, especially if takeovers are one-sided. Experienced executives leave, and the information systems they leave behind are no foolproof substitute for people. History indicates that the contraction of R&D teams and the dropping of projects lead to a decrease in pipeline for some companies.

A notable example would be that of GSK (GlaxoSmithKline) where the effort expended in merging caused a precipitous drop in projected project launches. The merger of the primarily UK-controlled Glaxo-Wellcome (GW) with the primarily U.S.-controlled SmithKline Beecham (SKB) introduced some personal acrimony into the merger soup. The compensation of the SKB U.S. chairman was much higher than that of GW's UK chairman. This may not have had anything to do with the reasons why the SKB chairman resisted the merger. The SKB chairman, Jan Leschy, is now, a Biotech life-science investor looking for late-stage, very promising projects of the kind SKB would never have out-licensed. The market also did not like the merger and the companies valuation lost $29 billion between them. GSK is still—at the time of writing—of significantly lower value.

The other significant downside is that a bigger company is a bigger target for any litigation. As well as acquiring the assets, one also acquires potential liabilities. Bayer was a hot merger or acquisition speculation target until the Baycol debacle of 2001. Now no one knows what will be the accumulated liabilities, even though Bayer has desperately tried to reach settlements—and has done so on class-action suits. But those who chose to fight Bayer individually are many, and one of them was just—but not justly?—awarded $3–5 million, making it truly uncertain how much Bayer has to reserve for all of these litigations, and thereby making it impossible to evaluate how much Bayer is worth. Pfizer, a few months after merging with Pharmacia-Upjohn, and reporting for the first time a $3–4 billion loss in a single quarter in the fall of 2003 has blamed part of it on liabilities with Pharmacia's growth hormone.

The downsides will not stop mergers, but, hopefully, they will make them more strategic rather than just in order to satisfy overt business ambitions and needs. Merging to develop complementary strengths and eradicate weaknesses is still a good strategy for companies that wish to combine resources and join—often rejoin—the ranks of the mega-companies. Some of the fallout has led to the birth of a niche Biotech industry that no longer wants to work on early discovery projects based on the science of its founders, but rather is founded mostly by former Pharma executives who try to rescue projects from the merger fallouts. The later the phase of these projects, the better it is. There is a focused drive to get them fast to the clinic and into the market. These companies are a very different kind of Biotech; they should be called development companies because they usually operate in clinical trial Phases I–III. They do, however, compete for funds with traditional Biotech, which, it should be remembered, started out by making biologicals before going on to provide the technologies to Big Pharma by leaving the capital-intensive stages, such as clinical development, to Big Pharma.

These development companies thrive because investors want faster return than the 6–10 years required when really starting at the beginning of the process. They are also predicated on the success of a few high-quality good projects that were dropped from Big Pharma because of merger politics. Finally, the investors believe, and they are led to believe, that founders have incredible insights into the value of the project at hand. If so, this benefit of a neglected or dropped project should belong to the stockholders of the merged companies. Yet it now will fall to the stockholders or investors of the newly formed companies that take over these projects for close to nothing. No Big Pharma executive will admit that he made a mistake, or that he might have eased out people with this "golden handshake." Most importantly, however, these golden egg projects are rare, but whatever can be licensed in from Big Pharma, it should be noted, was not pursued there because it was judged to be not up to the Big Pharma's standards. The judges of this in Big Pharma see collectively 100 times as many projects as the Biotech people, and are obliged to drop the least promising. In addition, these new companies taking on projects at later stages will need a lot of money because late-stage clinical development is expensive. It is much more expensive than financing a discovery-based Biotech start up. It works the same for a development company as for Big Pharma. The FDA does not allow shortcuts, and statistics will require the same number of patients for both types of companies. They are more likely to fail, but they will do so reasonably fast. Thus, they are fiscally interesting investment objects that now totally undercut the capital markets for the true Biotechs, whether struggling at early-stage discovery, or clinical development with a home-grown project.

The lure of getting something very valuable from Big Pharma for very little is going to persist, and it is also true that people only remember the successes. Repositioning is another strong Biotech investment trend. One tries to licence in from Big Pharma compounds, dropped from merger or other reasons, which were proven safe in Phase I, and develop them for an often smaller indication for which they believe the fit is better than it was for its original bigger indication. It may be good business for Biotech with a $200 million indication, but for Big Pharma dropping it would still be the right decision if the drug would not have worked in the intended $1 billion indication. Such successful repositioning stories are rare, but they are keeping VCs dreams alive.

Is bigger better? They have more money, they are bigger, but bigger doesn't mean they are better. Once you become so big, and you have so many research sites, there is so much knowledge in so many places that never gets to be used in the right place. The companies may be becoming too big and too top heavy to be properly managed. This too will only provide opportunities for small companies to grow. Although the quality of drugs has improved tremendously, the expectations of the stock market, patients, doctors, and government have grown even faster, which has created a gap.

GSK announced in 2002—after having built the grandest research center in the world—that it would break its research into Biotech-size units of 200 to 300 people in order to increase research productivity. These units have facilities for biology and chemistry, so they are similar to the Biotechs that now acquire their own chemistry of 80 to 100 people. It seems that Big Biotech and Big Pharma are testing the optimal size for research units.

For clinical development, there is the comparison with CROs as an external resource if you do not have the resources internally. The CROs act as external benchmarks for the industry. For marketing, there are no comparisons, only measures of muscle mass! Even if you are a small Biotech, you can enter agreements with the strongest Pharma, if you have something promising in their area of market strength. Co-marketing is a successful invention.

What Drives Mergers (Part 2)?

Objective analysis of mergers and takeovers by smaller Biotechs reveals something important. Examples range from acquiring companies for their technology or expertise, or their product and associated revenue streams. The interesting objective analysis is that while the business driver of this activity is textbook capitalism, the fiercest critics are the investors. There are many examples where the managers and executives are operating with a five-year plan, while the investors have a two-year horizon. Therefore, while hindsight may prove an acquisition to be insightful and profitable, the immediate response is often to be critical of the merger or acquisition and the stock price falls. The executives thus spend an inordinate amount of time—time better spent delivering the promises—defending their rationales for expansion in presentations to the investors. Of course, there are examples where hindsight indicates the costs of the acquisitions are too high, especially where the fit is obvious, but in general, the disconnect between the executives' and the investors' horizons causes real problems in Wall Street.

In case the point is lost when made in careful language, investors are the solution and the problem. Without them the companies wouldn't exist. But all the time investors work on their two-year gains in a world of five-year strategic plans, the pressures on companies to perform is counterproductive. Executives are distracted by the obligations of investor PR. This is probably true for all industries, but for the Pharma industry with product development cycles extending way beyond five-year plans, fluctuations in investment, as reflected in daily Wall Street valuations, demonstrate vividly the long-term futility of investors having a policy of "gain chasing" within a matrix of long-term plans. Perhaps it is a truism that stock markets are polite gambling for the elite. When the stock markets in Hong Kong suspended trading in the last recorded Black October, gambling on horse racing allegedly doubled. *Are investors the primary cause of high drug prices?*

chapter

26

WORKING WITH THE FOOD AND DRUG ADMINISTRATION (FDA)

EMBRACING THE FDA

Surprisingly many companies do not seem to take the FDA seriously enough, or they grossly overvalue its powers. Their employees have rather vague, general, and often fearful views of the FDA, and they do not learn in detail the workings of this important governmental organization, but leave it to their company's small group of regulatory experts. These experts act like high priests and interpret letters, phone calls, and meetings, and the atmosphere they generate for the rest of the company is "stratospherically cloudy," much like that generated by the Kremlinologists watching the Soviet elite from the outside. This is sad and counterproductive since the future of a company's products is dependent on approval by the FDA and the best thing a company can invest is time to understand the FDA with a view to improve all interactions. Understanding the FDA is necessary at all levels from heads of research, development, and marketing, down to sales staff at a trade show. Nevertheless somewhat scary letters from the FDA described below are almost never discussed outside of a tiny group within a company.

The FDA is set up to protect the public from potentially dangerous foods and drugs and to assist in the evaluation of the therapeutic benefits of new drugs using the best available science. Its sole mission is to protect the general public. It is specifically prohibited from controlling the actual practice of medicine. It is always being asked to do much more, such as regulate the efficacy and claims made for vitamins and food additives, but it resists attempts to expand the remit if only because it is impossible to test their effect.

The head of the FDA is a political appointment, but this should not compromise standards and procedures. Efficient processing at the FDA can make drugs more easily approved, and that would be unfair in its implication that less care is being exercised if a particular FDA commissioner oversees a higher rate of approvals during the tenure of an administration sympathetic to industry. Society demanded quicker approval for AIDS drugs, and pressure is on the FDA to regulate both efficiently and objectively.

In general, European approval of drugs is faster, and Japanese approval of drugs is slower than FDA. This is important because the patent life is "ticking away" during the testing and approval phases, so the effective patent protection and long-term earnings are very dependent on a fast approval process. Of course the 24-month trial will take 24 months, but if the FDA can make up its mind within a shorter rather than a longer time after the data are submitted, that may mean billions of dollars just on a single drug.

What the FDA Does

The origins of the regulation of drugs can be traced back to 1820, and the formation of a regulating body, the Bureau of Chemistry, to 1862. The Food, Drug and Insecticide Administration spun off in 1927 and became the FDA in 1930. The FDA is governed primarily by the Federal Food, Drug, and Cosmetic Act (1938), which has undergone many modifications and amendments such as the FDA Modernization Act of 1997. Its responsibilities include the following: Drugs—Prescription, Over-the-Counter, Generic (Center for Drug Evaluation and Research); Medical Devices—Pacemakers, Contact Lenses, Hearing Aids *and* Radiation-Emitting Products—Cell Phones, Lasers, Microwaves (both with the Center for Devices and Radiological Health); Biologics—Vaccines, Blood Products (Center for Biologics Evaluation and Research); Food—Foodborne Illness, Nutrition, Dietary Supplements *and* Cosmetics—Safety, Labeling (both through the Center for Food Safety and Applied Nutrition); Animal Feed and Drugs—Livestock, Pets (Center for Veterinary Medicine); and Combination Products—Drug-Device, Drug-Biologic, and Device-Biologic Products, such as drug delivery devices, and, for example, drug-cardiovascular stents (through the Office of Combination Products, established December 2002). It enforces a whole host of laws[148] set up to protect the American consumer and patient.

[148] See http://www.fda.gov/opacom/laws/

The FDA's mission statement is as follows: "The FDA is responsible for protecting the public health by assuring the safety, efficacy, and security of human and veterinary drugs, biological products, medical devices, our nation's food supply, cosmetics, and products that emit radiation. The FDA is also responsible for advancing the public health by helping to speed innovations that make medicines and foods more effective, safer, and more affordable; and helping the public get the accurate, science-based information they need to use medicines and foods to improve their health." At its most prosaic, it lists the 1,500 physicians that have been found guilty of unethical practice in clinical trials.

Through the activities of the then new—since November 2002—but since departed commissioner of Food and Drugs, Mark B. McClellan, M.D., Ph.D., the FDA had recently launched five major initiatives for Efficient Risk Management, Better Informed Consumers, Patient and Consumer Safety, Counterterrorism, and to build a "Strong FDA."

At the simplest level, drug companies seek FDA approval to do anything that involves the testing of therapies in humans. The FDA authors and publishes manuals and other publications[149] that establish the guidelines for "Best Clinical Practice" which insurers such as HMOs use to guide their own activity of reimbursement of medical bills, and so on.

The FDA professionals are organized according to specialties, but, even though it is a large agency, it cannot have all the expertise and it cannot have the latest in science and technology at its disposal. Thus, the FDA relies heavily on expert panels to evaluate new scientific evidence for or against a diagnostic tool, a therapeutic approach, or a risk evaluation formula. The FDA and the drug companies usually want the same experts. Who would not want the best and most up-to-date judgment? But the same academic expert may, as a consultant, encourage development and Proof of Principle testing of his or her idea at a Pharma company, and similarly caution about it to the FDA when counseling on another drug from a different company. Such turnabouts—based on expert opinion—are not uncommon, and are most often motivated by new insights and not by some dishonesty many ascribe to these experts of all kinds. But, of course, it is no fun for the companies when the goalposts are moved. FDA expert panels are an important driving force in clinical medicine and in drug development. The FDA is responsible to set the bar on safety and efficacy based upon "best available science" and it does so by convening regularly the best authorities from academia and clinic to review new findings, concepts, and methods. The results are modified safety and efficacy test requirements that will guide trials and the approval process in a given area of medicine.

---

[149] See http://www.fda.gov/opacom/7pubs.html

Many experts flaunt their FDA panel expert status, which means that they are often called to judge new diagnostic and treatment evaluation in a condition matched to their expertise. Companies like to have people who know how the arguments are laid out at the FDA and what are the issues to be addressed up front, as long as this leads to safer medicines faster and cheaper. So far this is the case.

The more *chronic* trial-regulating activities of the FDA can thus be leisurely learned and appreciated by drug companies. The FDA also has an *acute* set of actions that companies do not wish to learn the hard way. It is responsible for initiating and enforcing drug and food recalls, market withdrawals, and safety alerts. They have even recalled a particular water used in manufacturing drugs.

All of the FDA's correspondence is public and can be obtained on its web site. One of the worst letters for a Pharma executive—besides the letters from its board of directors (BOD) and major investors—to receive is a "warning letter." Threads of disputes can be followed from complaints to actions on http://www.fda.gov/foi/warning.htm. Most frustrating for company executives is that these complaints often start with the alleged misrepresentation of a drug's efficacy for too broad a range of indications without due reference to its contraindications made by the company's representatives at a trade show. The text may read something like the following:

> *The Division of Drug Marketing, Advertising, and Communications (DDMAC) has identified promotional activities that are in violation of the Federal Food, Drug, and Cosmetic Act (Act) and its implementing regulations. Specifically, representatives of [your company] made both false and misleading oral statements about [your drug] at [your] promotional exhibit booth at the [41st Interscience Conference on Antimicrobial Agents and Chemotherapy (ICAAC)] held in [Chicago, Illinois] in [December 2001].*[150]

The letter goes on to describe details of which particular false claims were made and, for example, alleges that representatives belittled the warnings on the label. It continues:

> *A fourth [company] representative also engaged in false or misleading promotional activities about the efficacy of [the drug]. Specifically, this representative stated that [the drug] "is approved for a broad indication" and characterized it as a "miracle drug."*

---

[150] See www.fda.gov/cder/warn/2002/10666.pdf

The letter then indicates how this error needs to be redressed.

> *[The company] should immediately cease making such violative statements and should cease the distribution or use of any promotional materials for [the drug] that contain the same or similar violative statements. [The company] should submit a written response to DDMAC on or before [date], describing its intent and plans to comply with the above. In its letter to DDMAC, [the company] should include the date on which this and other similarly violative materials were discontinued.*

The FDA could demand that the company write by registered letter to all participants at the congress to ensure that these incorrect statements are corrected and that new, potentially dangerous medical practice does not arise as a result of false information obtained by health professionals at the meeting. Much of this was "toothless" before serious fines were levied. The almost $0.5 billion fine Pfizer had to pay for promoting Neurontin for pain treatment without trial data certainly encourages some companies to be more careful; others regard this as a cost of doing business. Until one company's executives are held personally responsible, this will not change. Some civic groups argue that since broader marketing endangers people's lives by using a drug in an indication for which it was not tested and where it even may be harmful, those executives, it can be argued, should be personally liable.

Another matter receiving due attention from the FDA is the production of biologicals. It is a tempting yet specious argument that if approval has been given for a particular new chemical entity or biological, then it doesn't matter how it is produced. That is untrue. If a new synthesis is invented for the chemical, it has to be approved, as it may have new starting materials, or just new vendors for these chemicals, which might lead to a trace contamination present in the new product of as little as 1 part per billion. The drug is then different from the one originally tested and approved, and therefore needs to be verified safe. This does not mean that there were no trace contaminants in the originally approved drug at activities below today's analytical detection, but that the previous formulation had contaminants below what our bodies' enzymes can detect and have to metabolize. In the original studies, any contaminant was simply tested as an "ingredient" of the medicine and was found safe or inert; thus, the formulation with drug and other ingredients was "approved." No one likes to change vendors and synthesis routes for a successful drug. When it becomes generic, then of course new companies will synthesize it in their way and will seek FDA approval for this new product. These generics will not have to go into the expensive Phase II and III efficacy trials. Those established before for the drug and then proven during its patent life are taken to apply.

Biologicals have the same problems. For example, the manufacturing process that uses engineered bacteria to make erythropoietin (EPO) was once changed, and the efficacy and safety of the final product had been altered because of mutations occurring in the biological. This is not acceptable as the mutated EPO has not undergone clinical testing. These batches of EPO had to be destroyed.

The FDA is part of the system of medical practice prevalent in the United States of "evidence-based medicine." This means that there is objective evidence that a given therapy works. Within the execution of a more liberal practice of medicine, if a trial has shown that a particular beta-blocker works against a placebo, then physicians have the right to substitute a drug in the same class.

One area of medical research that is too large for companies to engage in is mass clinical trials, and even the FDA has difficulties in general to regulate, is the one of very large and very long-term clinical trials. Very large trials can be used to test theories about the benefit and harm of interactions between such factors as diets, aging, environment, lifestyle, disease prevalence, course, severity, and outcome, and the long-term effects of some already approved and very commonly used and often, but not always, cheap drugs. These fall outside the control of the drug companies and the FDA. The theory being tested might be, say, the effect of hormone replacement therapy (HRT) on cancers. This actual recent trial discovered long before its planned five-years duration that women on HRT had a disturbingly higher incidence of breast cancer, which was determined—quite rightly—to be much more dangerous than any protections it was also found to indicate. The perceived stock of HRT as a therapy plummeted, as did Wyeth's stock value. Only women who severely suffer from reduced hormone levels and insist on treatment are still receiving HRT, and drug companies have withdrawn their TV advertising campaigns encouraging women to consider HRT. In another large and long-term trial, it has been seen that ibuprofen treatment may provide very significant protection against breast cancer. Again, the evidence was so compelling that the results of the trial were published before the trial had been completed. Aspirin as an effective drug to lower thrombosis risk has emerged from a trial like this.

Where drug companies may suffer, and this is unavoidable, is whenever these large government-sponsored trials show that, for example, aspirin through its antithrombolitic activity is better than brain surgery and better than all the modern drugs designed specifically to be anticoagulants. Daily "baby doses"—81 mg—of aspirin are now taken as preventative medicine against stroke by many careful individuals. There is no benefit to drug companies in findings such as this. Acetylsalicylic acid made by Fredric Hoffmann for BAYER and tested on his father ($n = 1$, consent probably was given) was patented in 1887, so its patent life expired long ago.

One area where there will be developments is in the area of "combinational drug therapy." Currently, except where a drug is being developed as an adjunct therapy, drugs are tested by themselves and are compared to placebos and competitors. No one comes out with a new formulation comprising two or more new drugs. In practice, many medicines are used together, and patients have to take multiple pills at different rates and times of day and, for example, before, during, or after meals. It gets quite complicated. In response to the multiple antiviral therapy used in AIDS treatment and HIV infection, some foreign—notably Indian—manufacturers have produced single-dose pills combining the drugs developed by different pharmaceutical companies, and, according to some observers, breaking multiple patents simultaneously. It seems likely that for complex diseases, complex combined therapies will be invented and have to be regulated by the FDA.

The bottom line is that the FDA does a good job and all companies should treat it with respect and give it the attention it deserves, and not just as insurance against punitive action.

chapter

27

REGULATING REGULATORY REGIMENS RELIABLY

BALANCING SAFETY VS. EFFICACY

Are all withdrawn drugs taken off the market for only good reasons? The FDA is there primarily to protect the public. Protection comes from making sure that a drug is effective, but primarily to make sure that a drug is not dangerous or harmful when taken in the recommended doses. But the FDA is powerless to stop physicians from poorly prescribing drugs. And "ignoring-label" use by patients and physicians has led to numerous withdrawals, including that of Baycol, with severe effects on the company and the industry. Occasionally, as we have mentioned, drugs such as thalidomide, which can hardly be described as unequivocally safe, make their way back into the market because the condition they treat, in this case leprosy, is much worse than the side effect, in this case, of birth defects, which is less of a concern to the patient population with leprosy. Interestingly, thalidomide would never have been approved for treating morning sickness had all the studies in the 1960s been conducted rigorously (or by the standards on teratogenocity which were partly adopted because of the thalidomide tragedy), and its use for treating leprosy would never have been uncovered. We must wonder how many more drugs have been shelved by the industry believing them incapable of gaining FDA approval or for those compounds that would be approved because of the apprehension about possible litigation since companies are sure that physicians will prescribe them inappropriately and patients will take them even if they have established contraindications.

There is no established way whereby allegedly "unsafe" drugs can find their nonprimary indications, but "safe" ones find many indications. This is an important part of the business of the drug industry. To nonscientists it may be surprising that antiepileptics have a variety of uses especially in pain treatment and, less often, as mood stabilizers for people with bipolar—or manic-depressive—disorder.

The reason is that these neurological and psychiatric disorders are related to nerve activity, which is, in general terms, governed by the activity of ion channels in the nerve membranes and nerve terminals. Pain signals require pain fibers to "fire," and epilepsy occurs when populations of neurons fire in a way which recruits other neurons to fire, resulting in a seizure, which, depending on where it happens in the brain, can manifest itself as a characteristic motor seizure or other complex sensations or behaviors. The side effect of "suppressing seizures" is quite benign in the context of treating pain or mania. However, other side effects such as promoting lethargy or making one slow-witted or inducing increased appetite and weight gain would be unacceptable in certain contexts, especially if one is treating, for example, chronic pain. The fact that ion channels are so broadly distributed and are part of every biological function means that drugs that affect particular channels can have a broad utility, but their side effects may also be broadly manifested and potentially severe.

The use of several tricyclic antidepressants to manage pain may be additionally surprising. Occasionally, of course, slight modifications of the drug can make it more effective in treating the "side effect" and make it less active in treating the original condition. Modifications also help reinforce new patents on a new formulations use. As mentioned before, the discovery of Viagra as an impotence medicine is based on its embarrassing and almost missed side effect as belatedly discovered from the original cardiovascular indication trials.

No matter what is being treated, all physicians and patients should always remember that **every drug can be taken in doses where it will become harmful**. A drug may be considered safe when its side effects are relatively benign, can be predicted, and monitored. Drugs thought to be safe become decidedly unsafe whenever deaths become associated with a particular drug use. When these deaths are largely inexplicable—or so-called idiosyncratic—the drug is in severe trouble and the chance of its future use for anything is severely curtailed. Certainly, if there is any other drug for the same indication, then everyone will switch to that one.

These issues present real problems. On the one hand, drugs finding extended use are intuitively helpful for the patients and the manufacturers of the drugs. On the other hand, if drugs that are no longer under patent but generic find additional uses, then that will kill incentives for drug companies to work on that drug's newly found indication. Thus, the "best" general treatment for prevention of migraine attacks is considered to be the blood pressure reducing propranolol because it is cheap, not that it is especially effective. Do the anticoagulant properties of aspirin provide the best prophylactic treatment for stroke? Justifiably "Yes," even when compared with drugs that have been designed and approved for preventative stroke therapy. Already in 1869 it was pointed out that the then new aspirin was a dangerous drug because it promoted bleeding, but this risk factor or side effect has now been turned into a major disease-preventing factor. Do the as yet unproven but indicated anti-inflammatory properties of ibuprofen fight against Alzheimer disease (AD) and breast cancer (BC)? Possibly for AD and probably for BC.

Thus, we see that the lack of even a putative patent protection for an indication renders research into drug development for that indication unviable. Drug companies will not invest millions when aspirin or ibuprofen does just as good a job. Unless the U.S. marketing department of a pharmaceutical company backs a compound, it will not be available.

Is There Something Wrong with the Patent Law?

Most certainly! It is often said that patents cause higher prices. The often-stated evidence is that generic drugs—with no development costs—are cheaper. Many politicians and activists would seemingly like to go from indication to generic drug without going through the patent-protected phase exemplified by high prices. This is, we hope, clearly absurd. It is hardly ever stated that the high prices are almost guaranteed by the fears of future or impending expiration of patent protection as well as the risks of the period of market dominance being eroded by both direct and indirect competition, as well as the legitimate fear of drug withdrawal because of adverse effects or inappropriate off-label use and litigation. The price of specific drugs is also driven up because of the fear of the potential prospect of one of the *other* drugs in a company's portfolio being subject to litigation.

If one is seriously concerned about drug development for new indications, one should question whether patent protection shouldn't be extended and whether settlements and damages awarded pursuant to litigation should be based on actual damages rather than on the perceived wealth of a company (inflated as it is, of course, by high prices). It is a commonly held view that patents cause high prices. In contrast, it is more likely the projected loss of a patent that inflates prices. Might more of a free market with lifetime patent protection for a company's assets result in lower prices?

Drivers for higher prices are the desire of investors for a quick return, the threat of litigation, and history of drug failures. The other business factors—fear, caution, and obstacles—have been suitably emphasized throughout this section.

The other burning question is, how could changes to the FDA's *modus operandi* positively affect health care? There is no guarantee that changes will improve outcomes.

The FDA can and the European agency already does say that it will only approve three to four drugs for the same target, giving sufficient variation in side effect profiles such that every patient will find something they can tolerate. However, no action has been seen on this. When or if this happens, Big Pharma will not be interested in making numbers 5–10 drugs in a class and will likely place all its bets on more marketing power and raising the prices. But it will use its research machinery to look for other targets in its disease of choice. This might eventually, and hopefully, also benefit treatment-resistant patients within the patient group, or it will look into somewhat smaller diseases. One aspect of drug discovery that Big Pharma often neglect to predict and project is that if they find a drug for an indication, the prevalence of the indication may grow with increased market awareness. Which company executives earmarked male sexual dysfunction to have a broad audience in 1960 or 1980? Viagra has changed it.

part

IV

What Will Matter in Making Medicines?

chapter

28

The Hypothesis Is: There Is a Better Way

Proceed with Caution, but Proceed

Where do we go from here? Regulating against the pharmaceutical and biotechnology industries' successes is no way forward. Drugs are expensive, and this does not only apply to drugs that are from Big Pharma. All of Biotech's new biologicals are expensive, and the rational questions are: "How can we afford them?" and "How many new ones can we afford?" Reforms will be needed if society wants everyone to have access to drugs. Making them cheaper by legislative means will not make them affordable for the vast majority. Unless insurance—whether private, controlled, or governmental—covers the cost of the best available medicines, investment in the development of new drugs is almost guaranteed to decline. If a new drug is enormously expensive—as, for example, the three new biologicals against rheumatoid arthritis happen to be—should the drugs' use be limited by the depth of the patients' pockets, the physicians' judgment of medical need, the insurance companies' guidelines, the FDA and its equivalents, or by state or federal legislation? Who are the better—in the sense of most deserving—shareholders? Those of the drug companies or the insurance companies? Who should be lobbying for whom? And who should have the biggest lobbying vocal chords?

Later we shall attempt an analysis of each of the potential governors of medical practice. But, for now, a reminder of the parameters that govern drug discovery seems germane to the argument.

Drug Development Remains Part Science and Part Art

Scientists and other professionals in the industry are poor in predicting complex responses to drugs and the multiple drug interactions that may occur in one or a few of a million patients. As a direct result, drug development remains part science and part art. The first step in the artistic process is to determine the Proofs of Principle for drug efficacy and indication or disease therapy. Big Pharma companies spend millions to prove the principles of a new therapy—that is, show that the proposed drug target indeed can be hit by a selective drug causing the desired therapeutic effect and that this effect is robust enough to be of therapeutic relevance.[151] Can this cost be reduced in the future? Possibly, by establishing within science and with the FDA new ways of determining the principles in simpler studies.[152]

Incidentally, when a company spends these millions, the expenditures do go into the economy. Without embarking on a roller-coaster of an economic argument, society in general and segments of society, such as physicians and medical instrument industries and chemical laboratories, that rely on the pharmaceutical industry's expenses, benefit from the drug companies' often profligate expenditures. The costs make drugs more expensive, but these specific costs are not the root of the problem of expensive drugs.

---

[151] This can be done with a safe enough and selective enough drug that is far from perfect in terms of ease of administration, and the like, but good enough to test in patients the hypothesis and be a Proof of Principle that will strengthen those in the company with a weak heart to endure the additional expenses of $300–500 million, and years of new trials with the better "to-be-marketed-final-drug."

[152] R&D researchers in companies are not waiting to examine which drug interactions will occur *in vivo* (i.e., during a trial) but through established *in vitro* methods using human liver *microsomes* to test at the level of liver metabolism how two drugs that are likely to be used together will influence each other's metabolism. In addition, companies will now look preemptively at whether a candidate drug will cause "QT prolongation." QT prolongation in an electrocardiogram (EKG) has killed many drugs lately. It's a side effect of the drug on a key ion channel—the *HERG*—in heart muscle that can cause the unmasking of hidden arrhythmias in patients who never had known that they have this hidden heart condition.

The question society should ask of its legislators is: "How do we curtail the negative influence of litigious society that backs the individual—often at punitive levels exceeding the "crime" of unintended damage—to the detriment of society?" Lawyers protecting corporations are portrayed as evil personified, yet the threat of often unfair litigation from unscrupulous attorneys ostensibly protecting the disadvantaged is a major driving force for not only the high price of drugs but also for the often premature cessation of development of promising clinical candidates for presently nontreatable conditions within a drug company. Although arguments are often politically motivated, we hope that neither side of the political divide would disagree that litigation against corporations does not encourage lower prices and, in the long run, society pays for every plaintiff's victory. The funds Big Pharma now puts aside as a reserve specifically to pay damages cannot be used productively. These are large sums: billions of dollars at each of the large companies. The recent court and jury findings against Merck in the case of Vioxx shake not only the industry but will make an already very cautious FDA even more cautions.

Why Are We Missing So Many Targets and Treating So Few Diseases?

Society is spending, and industry is receiving as revenues, $300 billion, whichever way you wish to look at it. The $300 billion that goes on drugs is only 8 to 9% of the almost $4 trillion total spending on health care. Some 10–14% of the industry's revenues are invested by the drug companies on research. Perhaps this should be higher. There are trends in the industry to take advantage of new technical and scientific breakthroughs, such as high-throughput screening platforms, combinatorial chemistry, genomic research, and clinical sample collections, and their very introduction indicates a willingness to spend. In some years, spending was 17 to 21% in some of the top companies. However, the stock market analysts punish companies that spend on R&D to the detriment of dividends, and, in this way, they make R&D spending fairly uniform over the whole industry. It takes very strong leadership to go against them as Merck did some 20 years ago when it increased R&D spending while stock prices were falling.

The drug companies, as we have explained, also benefit from public monies spent on health research, which, in the United States, goes through the National Institutes of Health (NIH). Gobally, between $70 and

$100 billion is spent on health-related research, which is still only some 2% of health care costs. Many societies outside of the United States invest less than 1% of their health care costs into research, and this is regarded by experts as far too little. Society can ask shareholders of Pharma and Biotech stock to be less ambitious or greedy with respect to the industry's returns on investment, which has been higher than for many base industries such as cars, electrical appliances, and energy, but cutting the prices of drugs through legislation is not going to help the industry tackle all the diseases that are not yet treated. Society needs to ask itself whether a *less profitable* Pharma industry—and do not forget that Biotechs go bust much more often than they succeed—would be a *better* industry. The health care costs associated with malpractice insurance and legal fees and administration should come under just as much scrutiny as drug prices. Society should want to afford the drugs that do become available. Amateur sociologists should note that already 400 million people, that is, those living in the most successful industrial economies in the G8 countries—with the exception, of course, of the United States—basically subscribe to—and agree with—the idea of less litigation, greater negotiation of drug prices, but also larger markets through socialized medicine as a fundamental basis, with additional coverage paid voluntarily by individuals. The 2003 U.S. Medicare Drug Benefits legislation is moving on some levels in this direction.

Spending more money on research into drugs may not actually help, for, as we have explained, in some indications the shortage of validated targets prevents progress. With new genomic and proteomic approaches this should be a diminishing problem as more targets are being found, but validating them as viable drug targets will remain a stumbling block or a key to drug discovery, depending on your perspective. Science is still found wanting in some areas. The unprecedented speed with which anti-HIV drugs came to be developed shows that when research is focused on a narrow area, progress can be made at an extraordinary pace.

How can society encourage companies to focus on diseases that are presently unmanageable, especially life-threatening diseases? The problem remains that there are about 400 disease entities as listed by the FDA—not caused by known pathogens such as viruses or bacteria—yet only 50 of them are commercially attractive for Big Pharma operating under the idea that anything that is not potentially a $1 billion per year drug is just not worth developing. "Commercial viability" is estimated based on the assumption that for any disease entity companies might hope to be able to treat 30% of those who have the disease and break even on their expenditure, because the making of a drug can cost up to $500 million or more.

Legislating to reduce the price of drugs and allowing insurance companies not to pay for the best available treatment is not going to help. You would have to have yet more patients treated for each disease.

Rheumatoid arthritis (RA) again gives a good example. When Medicare legislation was being passed in December 2003, the stock prices of Big Pharma nose-dived, especially the stock of Pfizer and Johnson & Johnson. This was despite the legislation being ostensibly to allow access for more people to more drugs.

Newspaper articles[153] at the time were understandably critical of Big Pharma's "me-too" drugs that did not signify huge advances in treatment, but some also indicated that the very expensive drugs from Biotech were modern "me toos," which, in addition, were extremely expensive: approximately $20–30,000 per year compared with $5,000 per year for COX-2 inhibitors. In the example quoted for RA, where the first treatments that slow disease progression are biologicals (antibodies to TNFα) such as: Enbrel (etanercept—Wyeth out of Immunex/Amgen), Remicade (infliximab—Centocor/Johnson & Johnson) and Humira (adalimumab—Abbott), these were considered "similar drugs" by reporters. Yet there is differentiating evidence—including different side effects and different ease of administration and delivery—which will guide physicians according to each patient's condition, risks, and needs. Having three disease-modifying drugs, where the therapeutic norm was—and still is—just incomplete and unsatisfactory symptom relief, new to the market within two years is quite a breakthrough. They can hardly be considered to be "replacing or supplementing an inexpensive generic drug"—that is, the generic methotrexate, which has many toxic side effects and is not specific to the treatment of arthritis, though it can and will be continued to be used as a basic drug therapy in severe RA.

Unfortunately, in many countries drugs such as Enbrel, Remicade, and Humira, launched with $20,000 per year or more price tags, will not be available to patients because the national health systems are not adequately funded to provide the drugs unless the patients are already crippled by this chronic disease. In other words, the new drugs developed are not always being given to the people who need them most, whether through physicians' reluctance to prescribe an expensive drug, or the insurance companies' reluctance to pay, or the patients' inability to pay. In the United States, the FDA will determine the best medical practice guidelines that should oblige insurance companies to reimburse the costs if physicians determine that these drugs are best for the patients.

---

[153] See, for example, Andrew Pollack: "Biotech Industry Banks on Medicare Law for Help on Drug Bills," *New York Times*, December 1, 2003, p.C4.

The societal costs in both dollars and loss of productivity of having patients with arthritis becoming severely disabled before becoming eligible for treatment should put the cost of the drugs in perspective. In a free market, the fact that physicians have a choice of three possible new treatment regimens should help drive down prices, providing the drugs enter the system as reimbursable medications. If legislation is generated that equates these TNF antibodies to generic methotrexate, then research into even better RA drugs will simply not happen in the future. It is always true that if a disease or condition is already adequately treated, companies will not invest in research for new therapies unless there is extremely strong scientific evidence that a significant improvement can be achieved over existing therapies. In such cases, someone will take the risk and see if one can realize the medical potential and *charge* for it.

One other reason some diseases are not tackled by the industry is that companies are devoting their resources to invading their competitors' markets and copying their successes, since this is a safer bet than breaking new ground. But from a business point of view, one cannot stop the development of drugs such as Levitra and Cialis penetrating the Viagra market by promising, in the case of Cialis, prolonged time of efficacy. The industry needs positive incentives to devote resources to distract it into tackling diseases that are not yet treatable (i.e., where there are no easy targets and/or the financial returns are projected to be less). We can as yet only speculate as to what such incentives might comprise.

The industry will likely, of course, have to change some of its own guiding principles. The implementation of pharmacogenomic principles ultimately leading toward "individualized medicine" will mean that "big" indications would be broken up to smaller ones. While this would, on one hand, be detrimental to marketing, on the other hand it would improve response rates, and perhaps clinical trials could be smaller.

More on the Hunt for New Targets

New disease-specific, validated targets are needed. We have mentioned before that oncology is rich in targets. This is neither an accident nor a surprise. It is rich in targets because there is so much basic research. An estimated $10.7 billion is being spent annually on basic research by the National Cancer Institute (NCI) ($5.5 billion), Big Pharma ($4.0 billion), and Biotech ($1.2 billion). Oncology is probably the only area of research where Pharma does not outspend the public.

In addition, it is relatively easy to obtain relevant biological samples of surgically removed tumors that can be transcriptionally profiled or subjected to proteomic analysis according to the approaches of modern target discovery. The treatment opportunity is potentially great, despite the fact that many of the targets are not on the surface of cells and are thus harder to reach (see Figure 28.1 for more details). The real obstacle to developing cures or therapies is that oncology is split into many, many, small indications; that is, there are many tissue- or cell-specific tumor types. Some of the drugs that work in breast cancer do not work, for example, in spinal cord tumors or small cell lung carcinomas. This makes the commercial value of these very specific targets very limited. Even though, with present pricing, a successful treatment in the more frequent tumor types such as prostate, breast, and ovarian cancers would be a multibillion-dollar drug, a drug for head and neck cancer might be "only" a $500 million per year revenue generator.

Radiation and chemotherapy continue to be the "workhorses of oncological treatment" for surgically unapproachable tumors. By working on pathways that are related to loss of control of cell growth, one hopes to discover targets that may be as generic to different cancer types as radiation is.

In contrast, something like schizophrenia, where there is much lower basic research activity, and where it is difficult to obtain relevant biological samples from brains of patients—and we have to guess from imaging data and unsatisfactory animal models—has few new targets. But it is a large indication with a large treatment-resistant group, and so the commercial opportunity is great. The two most recently introduced drugs sell for more than $4 billion per year.

Of course, another reason why treatment for some diseases remains elusive is that the Pharma industry does suffer from risk aversion exacerbated by prior failures. After 30 failed trials, there are currently few and far between stroke trials.

Why Are So Many Candidates Lost: What Can We Do with Them?

There is something that society, legislators, the FDA, and drug companies can do. It would take a lot of negotiation, however, since it would rely on altruism rather than a pure profit motive. The selection criteria for clinical candidates are so tough that, within a drug company, even good candidates do not get a chance to fail; they are dropped prior to failure, sometimes even before any suspicion of prospective failure.

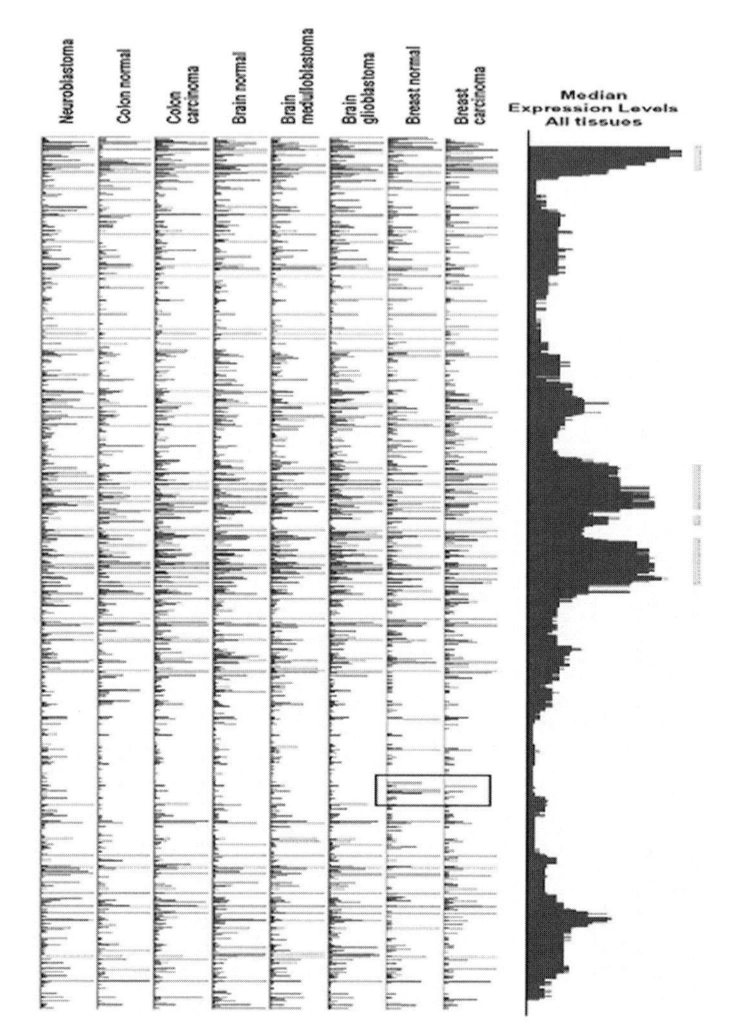

We touched on the value of tissue and serum samples from clinical trials that failed as well as from trials that succeeded. This is increasingly recognized by the companies as extremely important. The reason is so that scientists can do *transcriptional, proteomic, or polymorphism profiling* of tissues. Why do they want to? The primary aim of transcriptional profiling is to find different 'transcripts' within different tissues. In other words, while all genes are found in all the tissues of an individual, different genes—different transcripts of the genes—are active in different tissues, and in different cells in a given tissue, at different times, the products of these active genes are distributed differently in the tissues. An exception is found in the mammalian testis, which, for some reason has all the transcripts. The differential distribution of certain transcripts and gene products—proteins and thus targets—helps scientists make a drug selective to a specific target active in the selected tissue. If the tissue is a biopsy from a tumor, say, this means that one can target a protein specific to the tumor that isn't found in surrounding healthy tissue. This is why the most obvious application of this technology is in oncology.

Figure 28.1 Tissue Distribution of Transcripts—Prediction of Selectivity in Target Selection

Within all the drugs dropped by companies there are almost bound to be viable drugs for small indications. The industry could blunt its inherently competitive edge and work on ways to pool clinical candidates so that other companies can try them against their targets. One might not have to rescind all rights to the drugs so shared should they become successful in other companies' hands.

The most recent move by the most respected medical journals—which often can affect the popularity of a new drug extremely strongly—is to insist at the outset that we know about and document all the clinical trials, not just the successful ones, through a registry of all trials. If this directive succeeds (i.e., Pharma gives in), then we shall generate a knowledge pool that will help all Pharma in the long run. In this scenario, all patients and volunteers will have contributed to the treatment of a given condition. However, some litigious elements are currently pouncing on the notion that nondisclosure of failed trials in some way invalidates prescribing medications based on subsequent successful trials.

As a footnote on a chapter that has tackled complex issues of what can or should be done in the face of *very expensive* new drugs, *no* drugs for many indications, and a *massive* health budget, which makes, incidentally, the expenditure on patent-protected drugs almost insignificant, readers should pause to contemplate that a very large amount of money—some estimates are as high as 50%—is spent on the last few weeks of life. Access to drugs that can improve the quality of life over significant periods should not be restricted purely on fiscal grounds in an affluent society.

chapter

29

WHAT ARE "WE" ALL WORKING ON?

ALL LARGE COMPANIES WORK ON PSYCHIATRY AND NEUROLOGY

They also all work on cardiovascular treatments and therapies. (See Box 29.1 for breakdown of therapeutic areas.) But 23%—or $25 billion—of the U.S. pharmaceutical market is for neurological or psychiatric indications. You cannot be a big drug company in terms of total sales (except Merck) unless you have products and an R&D program in neurology and psychiatry. To reach the sales of Pfizer, Novartis, and Astra-Zeneca, you must have CNS drugs; you don't have a choice. Within this market there are the big indications: major depressive disorder (MDD), Alzheimer disease (AD), schizophrenia, and anxiety for which the revenues can be over $1 billion per drug per year. The middle tier indications in neurology and psychiatry attracting active pharmaceutical interest are bipolar—manic-depressive—disorder, Parkinson disease (PD), multiple sclerosis (MS), epilepsy, sleep disorders, stroke, and traumatic brain injury (TBI). More diverse and smaller indications, which are very difficult to treat for financial and technical reasons, would be narcolepsy, the ataxias (many different ambulatory or motion disorders), amyotrophic lateral sclerosis (ALS)—Lou Gehrig's disease or motor neuron disease—Huntington disease—formerly known as Huntington's chorea various neuropathies,

Box 29.1 Number of Top 20 Companies Designating Selected Therapeutic Areas as "Core" or "Additional" Areas of Research—1997

Therapeutic Area	Core	Additional
CNS	20	-
Cancer	20	-
Anti-infectives	18	1
Cardiovascular	18	0
Respiratory/allergy	15	1
Metabolic	14	2
Osteoporosis	10	1
Rheumatological/arthritis	7	3
Gastroenterology	7	2
Dermatological	5	4
Urological	4	2
Anesthesiology	3	0
Ophthalmological	2	0
Vaccines	2	0
Transplantation	2	0

This table indicates the number of the top 20 pharmaceutical companies engaged in particular areas of drug development. This has not changed from 1997 to 2001 and beyond. Twenty of the top selling companies have to be in CNS, and only 2 wish to continue in vaccines. The top 20 companies of 1997 are not the same as of 2001 or 2005, yet they pursue the same indications. The mergers did not change portfolios, but are driven by the needs of portfolios.

and tardive dyskinesia. There are an extraordinary number of neurological disorders, enough to fill many encyclopedias.[154] There are always difficulties with neurological disorders, even at the preclinical stage, because the drugs have to cross the blood–brain barrier to gain access to the brain, and it is very difficult to avoid cardiovascular or gastrointestinal side effects. The reason is that many drugs affect directly or indirectly ion channels in nerve membranes, yet there are as many ion channels in the heart and its innervation equally available or even more accessible to the medications. Moreover, one should be mindful that the largest selling CNS

[154] Of the c. 400 plus diseases, over 100 are neurological. There are many neurogenetic disorders; about 6 of the 70 volumes of the *Handbook of Clinical Neurology* focus on neurogenetic disorders. The new *Encyclopedia of Neurological Sciences*, which covers psychiatry, is 4 volumes.

drugs, SSRIs, block serotonin reuptake; yet the largest serotonin pools are not in the brain but in the gut. It's hard not to get side effects and adverse reactions when trying pharmacologically to reach the brain.

Searching for Drugable Targets to Cure or Slow Down Alzheimer Disease

Alzheimer disease (AD), the destructive aging disorder, came of age in the mid-1980s. First described a hundred years ago by Alois Alzheimer, AD has attained broad scientific interest—it is "claimed" by both neurologists and psychiatrists as belonging to their specialty—and major economical importance. The ever-increasing number of aged people in the developed world automatically increases market size by 4 to 7 million per year. Trials are long and challenging, but companies cannot afford to ignore AD since the potential for truly significant revenues and profits is too great. It is also a disease that society recognizes and wishes to be treated, definitely slowed down, and preferably cured. Since it is a neurodegenerative disease in today's research climate, the first aim is to slow progression, with a future goal to devise strategies to create or engineer some replacement for dead and dying neurons. Unlike muscle, liver, or kidney cells, nerve cells in the brain do not regenerate in substantial numbers, though stem cells can be cajoled by a cocktail of growth factors into becoming neurons. This is a process we as yet know too little about, but many encouraging results have uncovered processes to harness nerve regeneration for spinal cord injury, an admittedly simpler site than the brain.

There is strong motivation from society and Big Pharma to solve and control AD. Many of the aging population are at risk, and since the disease destroys one's own identity and memories, it is a recognizably awful disease that is in too many families' future. The search for drugs to combat this indication has lasted more than 15 years. Scientifically, every piece of physiological, pathophysiological, and genetic evidence is scrutinized for leads and targets.

The clues—often misleading—come from rigorous observation. The Alzheimer brain is invaded by so-called plaques and tangles that disrupt interneural communication. The amyloid plaques kill neurons and create a cholinergic deficit—and large deficits in other neurotransmitters as well—which manifests as a loss of memory and recognition of, for example, family members, and can cause behavioral problems, even violence. But the pathological and biochemical manifestations of the disease—seen mostly postmortem—from spontaneous and inherited mutations, probably occur late in the process, and earlier biochemical events have already committed the brain to become diseased. The genetic and biochemical evidence is very important, and there are ways to detect a propensity for the disease.[155] Intriguingly, though not universally accepted, AD is associated with inflammation, and some investigators are trying to demonstrate that nonsteroidal anti-inflammatories (NSAIDs) such as ibuprofen are an effective prophylactic against AD, something very difficult to prove.

Why is AD such a problem? In the United States there were 10 million sufferers in 2002, and this is likely to be 14 million by 2010. However, only 50% of the people with AD were being diagnosed in 2002, and given new tests and greater societal awareness, it is likely to be up to a 70% diagnosis rate by 2011. Only half of those diagnosed were treated—somewhat ineffectively—in 2002, and this again is likely to increase to around 70% in 2011. We have to ensure that the treatment is better, i.e., more efficacious. AD is treated by GPs, neurologists, internists, geriatricians, and psychiatrists, but the only drugs available now treat the symptoms; none is disease modifying. For further details, see Box 29.2.

Developments in Store for Other Psychiatric and Neurological Disorders

For major depressive disorder—the largest neuropsychiatric indication—it is and will continue to be hard to beat the SSRIs; they are safe, they are cheap (as of 2002), have a reasonable anxiolytic effect, and work well as GP drugs. They clearly are used as nondependence-evoking anxiolytics and even find prescriptions for sleep problems, although paradoxically one of the major problems with the archetypal SSRI, fluoxetine, is the *induction* of sleep disorders. It is an area where competitors such as paroxetine have claimed no effect on sleep as part of their marketing strategy. Some of their growing uses in adolescence are being questioned, but there is little that is left for these drugs to conquer in terms of age groups and disease categories extending to social phobia, in addition to

[155] For example, a Swedish double mutant of APP (Amyloid β A4-Protein Precursor) correlates with early onset and overproduction of amyloid peptide Aβ 1–40/42, and an ApoE4 isoform correlates with propensity for AD.

Box 29.2 CNS Drugs for Alzheimer Disease, Major Depressive Disorder, Parkinson Disease, and Multiple Sclerosis

The current drugs for the treatment of AD are the acetylcholine esterase (AChE) inhibitors—that is drugs that enhance levels of the neurotransmitter acetylcholine—rivastigmine (Excelon), tacrine (Cognex), galantamine (Reminyl) and donepezil (Aricept), which cause a small but clinically significant change in performance of memory tests (3.5 to 4.5 improvement on a scale of 70) and in tasks of everyday living. Memantine (Axura/Nameda), an anti-glutaminergic, has also now been approved in mild to moderate AD. The introduction of accepted diagnostic neuropsychological tests such as the MiniMental scale that are standardized over large groups of different ages and educational levels has given better tools for drug testing and thus development. The drugs that are needed are better "symptomatics" with both a higher efficacy than the current AChE inhibitors and with better tolerability through fewer gastrointestinal side effects. As well as trying to enhance memory, or stop it from deteriorating quite as quickly, the other noncognitive symptoms of insomnia, depression, agitation, and psychosis are another unmet medical need. Trials for such drugs may last 6 to 12 months. Trials for drugs that actually slow or halt or reverse disease progression would take 36 to 48 months.

One almost certainty is that the worldwide market for AD drugs will continue to increase up to $11 billion per year by 2010 or 150% up over the decade. What will the drugs be like? Now the dominant market share is held by AChE inhibitors, with the rest divided between antipsychotics, nootropics[156]—"smart drugs" or, preferably, drugs to make you smarter—and other diverse drugs for associated symptoms. The drug industry is focusing on diverse targets to combat AD, and the drugs being used in 2010 are likely to be β- & γ-secretase inhibitors, 5-HT_6 antagonists, acetylcholine receptor (nAChR $\alpha 7$) agonists, GABAa $\alpha 5$ inverse agonists, muscarinic receptor agonists, as well as more specific AChE inhibitors, and so on[157]—in other words, drugs that increase the availability or mimic the effect of acetylcholine, drugs that improve memory performance, and drugs that prevent the deposition of the amyloid plaques at the core of the pathology.

In addition, vaccination against the suspected culprit of the disease, the β-amyloid peptide, which aggregates into the plaques and which is formed by β- and γ-secretases, is being tested. The first vaccination studies against β-amyloid peptide were effective in removing some of the peptide from the brain in humans and in experimental animals—even reducing the number of plaques. But these studies have been stopped because of safety issues.

continued

[156] See http://nootropics.com/ for a general—albeit biased and simplistic—introduction.

[157] Other possibles: SAP kinase inhibitors, AMPAKine, NGF agonists, COX inhibitors, MAO-B inhibitors, mGluR enhancers and agonists, M1 and M2 agonists.

Box 29.2 CNS Drugs for Alzheimer Disease, Major Depressive Disorder, Parkinson Disease, and Multiple Sclerosis—cont'd

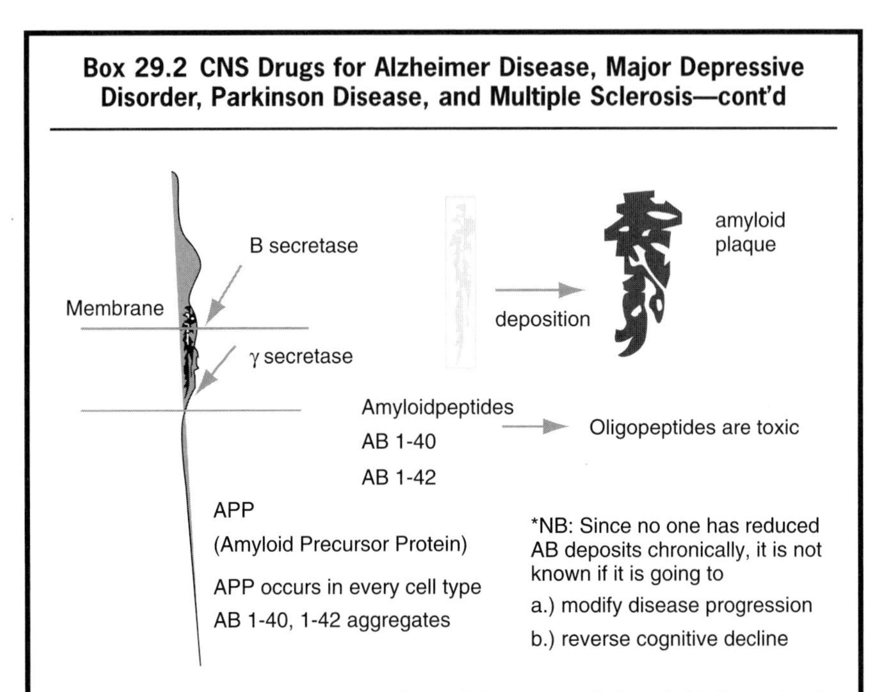

Putative Drug Targets in Alzheimer Disease and Amyloid Hypothesis

While there is no validated target for disease progression modification as yet, there is no shortage of possible targets. Merck, GSK, Pfizer, Novartis, Johnson & Johnson, Abbott, Astra-Zeneca, Taiho, and Boehringer-Ingelheim are all heavily invested in AD drug discovery. It is likely that others will emerge before the decade is over.

Their effectiveness in reducing amyloid peptide load is, however, so attractive that new vaccination strategies, using parts of the peptide as antigen, and other adjuvants to provide only a short-lived response, or the use of monoclonal antibodies manufactured to be used for passive immunization are being pursued. (See figure for possible targets in AD therapy.) The future drugs under investigation with potential in MDD include nonmonoaminergic transmitter systems, including several substance-P (NK-1) antagonists, corticotrophin releasing factor ($CRF_{1/2}$) antagonists, A_2A antagonists, and various ion channel blockers. The original therapeutic approaches to treatment of Parkinson disease (PD) was a model for neurotransmitter resubstitution and for precursor-type disease therapy, predicated on the loss of dopamine-producing neurons—found 60 years ago. L-DOPA, a precursor, a precursor, is given to enhance the effectiveness of the remaining neurons to

Box 29.2 CNS Drugs for Alzheimer Disease, Major Depressive Disorder, Parkinson Disease, and Multiple Sclerosis—*cont'd*

produce more dopamine, or ergot-type dopamine agonists are used to substitute for dopamine as agonist. Improvements of dopaminergic control are achieved by inhibitors of enzymes converting dopamine. But it is not expected that more dopaminergic drugs—L-DOPA, MAO-I, and COMT-I—are to be developed. More likely, spinoffs from other programs such as A_2A antagonists, NMDA agonists, and trophic factors to encourage cellular repair will be used in PD treatment. Surgical intervention to deposit either dopamine-producing cells or stimulating electrodes—such cell and tissue transplants first tried in 1984—have always provided tantalizing examples of extraordinary improvements. Research is continuing, with the most hope being for transplanting neuronal stem cells or their progenitors. There is likely to be a role for drug therapy associated with such attempts at surgical repair.

Disease progression is to be slowed by drugs blocking free-radical-induced neuronal cell death. Such compounds are developed to treat ischemic stroke as well as several neurodegenerative diseases, such as ALS, so any successful drugs will be widely employed beyond Parkinson disease treatment. For multiple sclerosis, new functional and radiologic measures of the brain lesions for detecting improvement are required. Establishing a Proof of Principle in animal models has been difficult since no adequate model of the disease mechanism to be addressed exists. Instead we replicate the symptoms with a widely used animal model—experimental allergic encephalitis (EAE)—which has been used for many years. Several efficient drugs in this animal model produced failures in the clinic, undermining the idea that this is a useful model of MS, and forcing companies either to abandon programs as they lack PoP possibility or to go to humans as soon as possible with a safe drug to look for efficacy. Immune suppressive therapy has recently been showing extraordinary promise in the treatment of MS (FTY 720 is in Phase II/III trials in MS).

major depression. Indeed, the effectiveness of SSRIs shows that serotonin function affects many disease states, that safe drugs find many indications, and that focused groups working on life-cycle management of a safe molecule can get it into new indications for the benefit of patients and manufacturers.

Since major depression is such a large and growing indication with a large treatment-resistant group in the millions of patients, new targets are being investigated. (Some details are provided in Box 29.2.) They share fundamental problems of the prevailing difficult trial possibilities in the United States with high placebo responses.

It is hard to show improved efficacy over existing medications, especially since the positive effects of antidepressants take so long to appear. Patients who are severely depressed and unresponsive to antidepressant therapy are still treated by electroconvulsive therapy (ECT). The goal is to find targets whose targeting will provide a fast onset of antidepressant effects and in particular a rapid reduction in the suicide risk.

It is not always good news if a drug is approved for treatment. Any drug slowing disease progression is, of course, very welcome. But the establishment of β-interferon as treatment for multiple sclerosis (MS) does make it more difficult to develop new drugs, which can no longer be tested against placebo, but have to be tested against β-interferon, thereby increasing the cost and complexity of difficult clinical trials. Clinical trials are particularly difficult in MS because the endpoints were not easily agreed upon. Since the disease pattern is characterized by relapses imposed on a decline of function, drugs that change relapse frequency, and drugs that stabilize the disease state or slow disease progression are difficult to differentiate between. Besides, since improvements are often part of the disease cycle and difficult to project, showing any modest improvement is difficult. Regrettably, no drug is so effective that its benefits can be shown unequivocally; they all rely on statistics. However, the recent acceptance of brain lesion imaging combined with clinician/patient reported relapses over time provides more objective criteria to evaluate MS drugs.

CNS Drug Development Is Still Fragmented

The major point to make is that there is much to do in CNS drug development. And all the significant, established drug companies are in the hunt for new treatments. The market is still fragmented, with no company having more than 15% of the market. Lilly, Roche, GSK, Pfizer, Novartis, Biogen, Bayer, and Sanofi-Aventis are all committed. Merck with 400 committed scientists has invested more than 10 years of research but has no product yet. Wyeth has also established a significant neuropharmacology R&D group.

Pain, neurology, and psychiatry are interesting areas with an increasing number of patients as society ages. None of the diseases causing nervous system degeneration can be cured today. In addition, there is growing understanding that psychiatric illnesses such as major depression, if untreated, but perhaps to some extent even when treated, is containing a neurodegenerative component—making the will to treat it greater.

We have in each of the major indications effective drugs that are not based on the full understanding of the disease mechanism. After all, the discovery of antipsychotic drugs such as dopamine D_2 antagonists was serendipitous. New understanding will lead to new therapies. There is no area that is standing to benefit more in the long run from the genetic and genomic studies than the area of diseases of the nervous system. But the fruits of the labor of discovery may force companies into smaller or even much smaller disease indications. In addition, many small companies are working on a single or few indications in neurology and psychiatry, but they will not be able to afford clinical trials in these indications (with the possible exceptions of acute pain, panic, or anxiety), so there will be a trend of licensing small companies' clinical candidates to Big Pharma.

PHARMACOGENOMICS IN SOCIETY[158]

The natural trend in drug development will be toward indications of lower incidence, prevalence, and market value. There will be no choice. In addition, attention will have to be increasingly given to "individualized medicine." But genomics-based, individualized medicine assumes huge changes in society's attitudes. This will become an enormously interesting debate.

The genetics of the population affected by disease and the pharmacology of current and prospective treatment is a major component for consideration in drug development. In this sense, pharmacogenomics—the molecular study of genetic factors that determine drug efficacy and toxicity—has always been part of decision making and project selection in drug development. The practice of complementing epidemiologic, diagnostic, and treatment-rate data with pharmacogenetics data was introduced in the 1950s, and the term *pharmacogenetics* was coined by Arno Motulsky in 1957, who wrote: "Genetically controlled drug reactions not only are of practical significance, but may be considered pertinent models for demonstrating the interaction of heredity and environment in the pathogenesis of disease."

Probably about $20 billion over the past six years has been spent on the use of pharmacogenomics in clinical trials and in preclinical drug

[158] Adapted from and reproduced with permission from T. Bartfai, *Pharmacogenomics in Drug Development: Societal and Technical Aspects* Pharmacogenomics J. **4**, 226–32 (2004).

discovery. The impact of these efforts on drug discovery process and pipelines is discussed primarily in terms of return on investment.

The Broader Picture

The political, social, and economic consequences of pharmacogenomics, such as the genotyping of larger patient populations in major diseases are also being discussed, albeit too little. Academic society tends to regard pharmacogenomics as a MUST for better or more effective drug development, but is leaving the discussion of the methods and results of pharmacogenomics to the Pharma industry, regulatory agencies, and, to a lesser extent, to academic researchers. This is a dangerous route because by the time we have genotyped people for 10 large trials in major categories of cardiovascular, metabolic, and mental diseases, thousands or even millions of people would have been genotyped, with much wider consequences for society as a whole, than are presently discussed or intended.

This is notwithstanding the insightful, conscientious, balanced, and courageous efforts of the FDA and European regulatory agencies to spearhead discussions with the Pharma industry on the use of pharmacogenomics. Some of these discussions are aimed at ensuring that the science is used properly, that biostatistics of enriched populations remain properly powered, and that not too many comparisons are made using the same populations—leading to a reduction of the power of the studies—as we see too often happen. The scientific foundations are solid and robust and will presumably be successful, and science corrects itself where and when needed.

Genotyping of a whole nation, as was envisaged and is ongoing in Iceland for the entire population of 290,000, has political risks. These risks were understood and carefully considered when the Icelandic parliament agreed to the DeCode project. What if a major vulnerability for a microbial or chemical agent were more common in one ethnic group (here in one isolated nation)? In the extreme, it could become an obvious and exploitable threat of the development of a specific genocide weapon. Luckily, even isolated populations are sufficiently outbred to make this risk small, though not zero, at the level of today's molecular genetics. In many countries, elite forces or sometimes the majority of entire armed forces are recruited from a single ethnic group, which would make them potentially vulnerable if massive genotyping turns up some vulnerability. Individual leaders are now safeguarding with secrecy their own DNA to prevent predictions of disease and so on that could affect political stability.

One may claim that this issue is not new, that we have known for many years about HLA-subtype differences between races and their consequences in disease responses: massive genotyping of populations to predict the safety and efficacy of agents will turn this knowledge from a blunt instrument to a sharp one. This trend to broaden genotypic knowledge about individuals in populations has to be carefully monitored at the political levels of society.

Social-Ethical Aspects

The broader ethical aspects of pharmacogenomics are only just being appreciated. Few drugs, indeed, too few drugs, are marketed today that require genotyping, although the clear coupling between the success of breast cancer treatment with Herceptin and BRCA1,2 genotypes provides a clear example where genotyping is not only recommended but is indeed required for the commencement of treatment. This example of Herceptin use based on patient selection by genotyping is, however, not going to be representative of the ethical challenges of the wide use of pharmacogenomics because (1) breast cancer is a life-threatening disease that is socially and sympathetically recognized and (2) no discussion of allele frequency, and the like among ethnic groups has surfaced as being significant or relevant. This will not be true for all diseases.

When similar requirements of diagnostic genotyping emerge for complex major diseases that do have ethnic or life-style components, much more ethical debate will occur. We should already recognize that by making drugs tailored for specific genotypes, we shall in fact precipitate these societal developments.

The capacity and ability to choose populations for which we develop the new generation of safer and, hopefully, more efficacious drugs are going to meet significant, appropriate, and rightful resistance as the "right to health" is embraced as universal. The focused development of drugs for a particular group of patients may not be societally or politically acceptable, unless the genotype is demonstrably more or less evenly occurring across different ethnic, social groups. The scientific underpinnings of the projects (i.e., clear coupling between genotype and drug response) are being taken for granted, but this indeed is as yet unproven and must be fully explored and validated between the Pharma industry and regulatory agencies before it should become dogma.

Social consequences may replace the already known and terrible global consequences of the economic imperatives that companies must obey. Today, whole continents are written off as a market for which a

Pharma company should develop drugs. With the genotyping of richer nations, such arguments may arise regarding some groups within a country. What should a government do if it turns out that a group that is genotypically defined, but because of its economical, political, and/or social state, is not judged as a promising market, and simply no drugs, safe enough and efficacious enough, are developed for this group? Should the free market permit an Africa-like development in terms of drugs being tailored, and costs underwritten, for a specific population within the country?

The Pharma Industry Dilemma

There is a great interest in increasing the chance that the drug companies' produce will have a higher likelihood of reaching the market, and staying there as safe and efficacious, than is presently the case. The Pharma industry is in a seemingly unrecognized or, at least, underappreciated dilemma. Pharmacogenomics promises better treatment response, at a probably higher drug price, but to the smaller patient groups who genotypically qualify.

If it turns out that a drug is developed without regard to genotype, but analysis of trial data shows that it is safer or more efficacious in a genotypically enriched group, we have no ethical issues to consider. It is good scientific and medical practice to learn from the data at hand and to use it for safer and better treatment of the next patient. For the companies—although the financial rewards may be different from the originally expected and projected ones—it is still better to restrict a drug whose development, premarketing, and marketing costs are already fully paid, than to drop it as unsafe.

Major Simultaneous Large Changes Are Required

For the expanding use of pharmacogenomics in Pharma decision making and drug development to become established practice, several parallel developments must take place in our society, of which the Pharma industry itself is only one factor. If these changes do not take place within a regulated, transparent framework, we will all be losers. The industry will not utilize the best available scientific data to create genotypically targeted drugs and, thereby, split large indications into smaller ones, if patients are not willing to genotype themselves. In addition, patients have no incentive to genotype themselves if there is not a concrete promise of a better, safer treatment and if there is a concomitant risk of being penalized by employers and/or insurers.

Our present inability to predict disease onset, severity, or treatment outcome—with the exception so far of a few cases with only small studies—turns the genotype data of a patient into a blunt, nonscientific, and potentially arbitrary instrument of discrimination in the hands of insurance companies, employers, and others. Once the data are generated, it will be much harder to create legal protection when an insurance company subsequently requests the data whenever such data were generated. Such legislation needs to be watertight and bulletproof. It is too easy for companies to discriminate indirectly through elevated pricing and deepening discounts for the "genotypically favored." Encryption to protect individuals has been well thought through in Iceland in the DeCode project—and even if the encryption is imperfect and breaks, the Icelandic socialized medicine platform will take care of everyone independent of genetic vulnerability and of our skills to deduce actual risk from such vulnerability data. Such a "guarantee" does not exist in countries where medical insurance is dominated by private companies.

The changes in strategic and marketing requirements, and in clinical trial strategies, will be profound once a large part of the companies' drug development projects are directed toward genotypically enriched groups of patients. Traditionally, the companies looked at what could be learned from a small group of patients in a familiar type of disease, in the hope that the lessons learned would enable them to make a drug for the much larger sporadic cases. With this in mind, the trials were almost always carried out in the large nonfamiliar case populations. This would change if the companies were able and were to make drugs so selective that they were only safe and efficacious in the genotypically specified group. Although we have seen this as being doable for some therapeutic antibodies, for small molecule drugs, this may be a much taller order. If, however, very specific drugs were to be made, then drug companies must take into serious consideration that if drugs are subsequently prescribed outside the drug's intended patient population as defined by pharmacogenomics, the drug may have to be withdrawn from the entire population if there turns out to be dangerous or fatal consequences. Off-label use and tacit acknowledgment of it may come to be revisited.

The regulatory agencies have so far been on top of these developments. Their dilemma too is that while they can regulate the approval of drugs from trials adjusted to smaller "genotypically enriched" populations, they cannot adequately regulate and control off-label use in clinical practice.

There are a growing number of excellent articles and reviews on the use of pharmacogenomics in the selection of clinical candidates from a pharmacotoxicological or toxicogenomic point of view, as well as on the use of pharmacogenomics in the design and execution of early and late-phase clinical trials. In addition, some authors have addressed the broader issue of the desirability, and potential risks, of stratifying patients for trials in order to enhance the likelihood of showing efficacy. The pros and cons will become clearer as clinical trials embrace more pharmacogenomic data, and as the data on SNPs, linkages, and drug response phenotypes become available more broadly. These data will then mesh with results in real clinical practice where the robustness of coupling between genotype and drug response in patients, after chronic, repeated treatment, can be evaluated.

Huge changes are needed for a wider application of pharmacogenomics in drug development; and such changes are slow, costly, and hard to predict in terms of social and political and even economic consequences. On the other hand, scientific trends continue with great momentum, and the large investments fueling them are guaranteeing that the case for pharmacogenomics assumes an irresistible inevitability. But, as is customary in the industry as well as in medical practice, a cost/benefit or risk/benefit analysis is worth carrying out.

Pharmacogenomics' Contribution to Drug Discovery, 2003

While examining the impact of pharmacogenomics in 2003, the most positive contribution of this field is found in the post-trial stratification of patients to enhance the safety of drugs where the gains are substantial for patients and companies alike. Although the use of large-scale genotyping carried out on patient samples collected during trials is becoming an ethical and moral issue, it is, at the moment, not a legal issue. Currently, these samples are owned by the companies, but do they truly own the data they may glean by examining massive numbers of SNPs? No matter how noble the company's cause might be, if there were no scientific hypothesis, agreed on by others independent from the company, then what is being looked at is likely to relate to the disease or to the drug response and would not be broader than that. Therefore, there should be some obligation for companies to disclose their findings without compromising their intellectual property rights.

The bottlenecks in drug discovery vary with the therapeutic area. For example, oncology, virology, and endocrinology are considered to be rich, and psychiatry is relatively poor, in drug targets. There are, however, many more drug targets in oncology or in virology that involve, for example, protein–protein or protein–DNA interactions for which the industry's present chemical libraries have few hits to offer; hence the starting points for medicinal chemists are limited. Consequently, people hope that genotyping in diseases where animal models do not exist or have proven of limited validity (e.g., schizophrenia) will help to provide new drug targets, and that such targets will provide completely new insights into the etiology and pathogenesis of these diseases. In such a case, the target-poor therapeutic areas would benefit. While no one really expected pharmacogenomics directly to assist with new chemical hits, this is hoped to come from better, more diverse, and bigger chemical libraries (a recognition now held by genomics-based Biotechs, all of which have bought or developed a medicinal chemistry unit to become just like traditional Pharma—an interesting and sustained trend).

It was hoped that, in the context of selection and validation of drug targets, pharmacogenomics would mean a huge breakthrough. This expectation by both Pharma companies and the public can be traced to the founding of projects, Biotech companies, Biotech–Big Pharma consortia (DeCode, Celera, CuraGen, Avonex, etc.), all promising that new drug targets and, thus, hopefully, new drugs, will be discovered for large indications such as obesity, rheumatoid arthritis, and schizophrenia. Indeed, in a 2003 interview the founder and chairman of DeCode, Kari Stefnasson, claimed that his company discovered disease genes in more than 20 diseases by examining the whole Icelandic population.

In the area of target discovery, the present drug discovery pipelines have, however, little to show that is based directly on human pharmacogenomics: only one drug approved in 2001 and two in 2002 can be traced back to pharmacogenomic-derived targets. This, of course, may in part reflect the long timelines of drug development. This is in spite of very notable examples of scientific knowledge breakthroughs from linkage studies and positional cloning, such as the recognition of specific channelopathies and, thus, the identification of subtypes of ion channels associated with mostly single-gene, genetic diseases. There are other examples in inflammation, oncology, and metabolic disease areas, yet one has to say that the majority of targets today addressed by Pharma research projects come from the painstaking work on signaling pathways. The use in this work of transgenic animals, most often of null-mutation carrying animals, has been the most commonly used tool accepted and adopted by the Pharma industry for the purposes of target validation and Proof of Principle in animal studies.

The large number of human SNPs has so far seldom shown the way to a new drug target, but this may change as we learn about polygenic complex diseases.

Human linkage data have led to a new focus on pathways and thus indirectly to new targets. Notable examples in the area of central nervous system (CNS) drugs include the ApoE 4/4 effect on the onset of Alzheimer disease, a now 10-year-old observation that has been repeated in numerous studies but has not, incidentally, provided drug targets but indeed very good ideas for research in the broader area. The more recent identification of neuregulin-1–schizophrenia linkage discovered in Iceland and confirmed in another population in Scotland has resulted in renewed interest in the pathway for this trophic factor signaling. However, it seems clear that neuregulin-1, or even its receptor, is not yet embraced as a drug target.

Whether target identification and validation come from targeted null-mutation, or random mutagenesis, such as is employed in the many forward mutagenesis programs around the world is not important for those in target selection. It is the appearance or the unmasking of a robust phenotype associated with the mutation that will be decisive in the early drug development decision, not how the mutation has arisen, be it naturally, targeted, or random.

Today, few drug development programs in the Pharma industry do not utilize transgenic animals with null-mutation or overexpression of the drug target as part of the development process. This part of molecular genetics has been fully integrated into drug discovery, together with the use of antisense technologies for target validation in the absence of null-mutation-carrying transgenics.

Urgency of Wider Ethical Debate on Pharmacogenomics in Drug Development

Pharmacogenomics is here to stay and provides a useful contribution to the development of new drugs in diseases for which we have no medicines, and, for other diseases, new, safer, and more efficacious drugs. This, hopefully, will justify the already made large financial investments and will also justify the enormous efforts needed by many players in society to limit its potentially harmful uses outside of the narrower, well-regulated area of drug therapy.

To be more certain than we can be today of this development, we need a forum to watch the early steps of ethical uses of the large-scale genotyping portion of pharmacogenomics. As outlined above, legislative, regulatory, and societal (insurance, employer) changes will be needed. Thus, the group overseeing these changes must reflect the political will of the countries involved and have experts as their aids rather than having expert committees without a real chance of their views being adopted into public policy. The Icelandic example of parliamentary commission and expert panels has worked well and is recommended for adoption elsewhere. It is the duty of the regulatory agencies and of the scientists to request from government the formation of a body that will broadly represent the varied views and interests pharmacogenomics affects. Industry has rarely initiated regulatory moves.

Meanwhile, the small-scale, individual, and disease-related genotyping is becoming part of the practice of modern medicine, to the clear benefit of everyone.

chapter

30

MORE TABLETS TAKEN PER DAY THAN MEALS SERVED: CHANGING THE INDUSTRIAL AND LEGISLATIVE STATUS QUO

COULD THIS HELP?

It is very easy to embrace legislative ways to regulate or deregulate the Pharma industry, but would it really help? If the aim is to try and get more and better drugs available to treat more diseases, how can politicians and lobbyists be expert enough to invent legislation that is supposed to improve Pharma productivity in the long run? Given the development times for drugs of about 4 to 10 years, it would take at least 5 to 12 years to find out if there had been an improvement and about 20 years to reverse the effect if it was determined that the legislation had made things worse.

No legislation can help directly, only indirectly, through creating the scientific and educational basis and the economical conditions for the deployment of new drugs, from development through approval, and then to the acceptance, prescribing, and purchasing by the market of physicians, patients, and insurance companies. In terms of expertise, the FDA working with the drug industry is preferable to congressional committees—and their international equivalents—if only because of the highly complex and specialized nature of drug development. However, when it comes to creating and regulating the market, only legislation can have major effects. In the United States there are more tablets used per day than meals served, so it is everybody's business, and thus it can only be politically and legislatively administered and "solved."

The judicial branch of government could help to shape a more risk-taking atmosphere. The continual fear of lawsuits against companies whose products have done unintentional harm should not be a reason for abandoning promising drug programs. Of real concern is that an administrative failure to create a favorable environment for a more creative and innovative approach means that too much of the energy of Pharma is channeled into "safe" projects. Efforts to make the number 3, 4, or 5 drug in a class drives up everybody's marketing costs and makes poor use of precious talent in reinventing the wheel, instead of having a go at untreated medical problems. But, primarily, improvements will come from a greater understanding between all parties, their motives, and their concerns. Legislation should only work to liberate—not distract—the creative impulses that drive drug discovery. All victories by lobbyists against other lobbyists are Pyrrhic.

Should Society and the Law Have a More Compassionate View of Pharma? Should Patent Laws Be Changed?

There are significant arguments about the effect of patent law. The United States has a special regard for intellectual property (IP) rights that are well protected in law. Removing or reducing protection would clearly destroy the business proposition. The World Trade Organization (WTO) is watching over patent rights with an increasing global reach to protect corporations; other less powerful nongovernmental organizations (NGOs) are trying to represent largely disadvantaged patients. Yet breaking the patent rights, when it happens, is done by government decree—such as is happening to treat AIDS in Southern Africa and in Brazil, especially. The U.S. military has also used the legitimate excuse of national security, and on two recent occasions has taken measures to produce smallpox vaccine and ciprofloxacin (Bayer) against anthrax when commercial sources are unable—according to U.S. government sources—to cover projected needs.

There are proposals worthy of consideration such as relating patent life to trial length since a long trial—or having to repeat a trial because of unforeseen circumstances—eats into patent life. Extensions could be granted for a variety of currently "unconventional" reasons that would encourage companies to look at indications for which it would take a long time to establish efficacy. Perhaps congress could add to the FDA's list of concerns and responsibilities, and charge the FDA to take this into account and advise, recommend, or even decide on prolonged patent protection where trial and approval are very long. Deciding on patent length from date of approval would be a relatively simple modification.

One area where the FDA might have made a difference had it been in its charter would have been in the area of cooperative working with other national and international authorities having the same purview. As mentioned earlier, over 800 compounds are sold in Europe, which are highly efficacious and cover some current and unmet medical needs in the United States but which will never be registered in America, because by now the patent life is too short to navigate U.S. trials and seek FDA approval. If the FDA were allowed to have been proactive in cooperating with the drug companies and the European authorities, these drugs would not have been restricted by the decisions of a few marketing executives in pharmaceutical companies. However, in view of the somewhat critical review[159] of especially the Canadian drug safety standards by the U.S. representative during a 2003 G8 summit, it is hard to imagine that this will happen anytime soon. This is a simple example of politics getting in the way of judgment.[160]

Rather than interfere with robust patent laws that have proved their worth many times, one approach that might improve the chances of development of drugs for indications that are not being adequately investigated or treated would be to extend the "orphan drug" status. Diseases with poor commercial outlook for drug development could be given special status to encourage investment in drug development.

---

[159] Also known as "unjustified bad mouthing," especially since Canada is highly organized and developed and its medical standards are comparable to those of the U.S.

[160] The FDA cannot force a European company to register its drugs in the United States but in the right circumstances there might be a way to make the FDA encourage such applications.

There are ways we could make drug trials cheaper. For example, the FDA has on file at least 30,000 measurements of changes in blood pressure upon the use of a β-blocker in a healthy and in hypertonic population. Every β-blocker, ACE inhibitor, AT-1 antagonist, and some diuretics were run against such controls to show that these new drugs in trial are effective in lowering blood pressure. Companies could use these extensive "historical data," and, save having to repeat one arm of a study. Since such a study would be very large and expensive, it represents a lot of money.[161]

Anything that can improve the chances of safe drugs being approved should be encouraged in a situation where current medications and therapies are not particularly effective in the fullest sense of the word.

Should Litigation Be Restricted?

The U.S. Supreme Court in June 2004 imposed restricted litigation against Health Management Organizations (HMOs). Could further restrictions on the extent of permissible litigation be beneficial?

There is no such thing as a safe drug. Drugs are created to effect cures or reduce symptoms that would be worse if ignored. As such, society should not be surprised if certain patients suffer adverse effects. However, defending the right to litigate—and demanding unlimited sums—against companies that have not intended harm is damaging society at large. From a business point of view, the problem is that the size of claims is not limited, so companies cannot reasonably calculate risk. Lawsuits—or just the threat of lawsuits—hinder drug development, prevent drugs from being utilized, stop certain indications from being investigated for pharmaceutical treatment, and have been known to cause drug withdrawals. It is incalculable to estimate how much better spent money would be on investing in drug development than defending often unjustified lawsuits. Lawsuits may not only cause drug companies to lose money as much as reduce the availability of drugs and increase the price of especially new drugs as companies reserve revenues to defend possible suits.

[161] Note that one notion contrary to that mentioned earlier, while companies, could perhaps save money and prove worth more easily against a placebo rather than the gold standard, some companies for selected drugs and conditions often themselves want to run against the gold standard. It will help marketing if it works. Taking a calculated dare in Phase III, might be better than waiting for a Phase IV challenge. Others feel differently about risks. It is on a case- by-case basis that the FDA allows you to run against an active drug or against placebo.

No serious defender of patients' rights would see merit in television advertising—such as was aired in December 2003—which allegedly states—somewhat paraphrased—"If you or someone you know has attempted or committed suicide while taking Neurontin call 1-800-LAW-FIRM." As we mentioned earlier, Neurontin (gabapentin) is an often-used antiepileptic that happens to be one of the most effective painkillers. Indeed, it might be considered the gold standard in pain therapy today. We make no claims about its appropriateness or its safety, but we do question the motives of such a modern manifestation of "ambulance chasing."

No one wants to restrict rights, but it would not be unfair to consider amending the instructions to juries to relate awards to the actual damage caused rather than the perceived worth of the company. Europe in general has taken an approach so that many lawsuits are thrown out of court, and many have a cap on damages. Lawyers there are prohibited from soliciting cases. While companies' sales income for the same drug may be in Europe vs the United States 1:3, the liability is 1:20 or higher. Ultimately, this trend may lead to a situation where Americans will not have access to some new but "tricky to administer" drugs. This may be good or may be bad, but we have seen what happens when Americans go abroad for treatment: it becomes selectively available and, in general, will be more expensive even for those who can afford it. Of course, the much higher prices in the United States compared to Europe provide the good balance for the larger litigation risk. But these trends in society are strongly affecting each other, and mix strongly with politics, as evinced by the controversy and concomitantly mixed and contradictory reviews over the new drug benefit legislation of December 2003.

Creative Lateral Thinking?

One area that might be in the realm of fantasy in this highly competitive IP-driven industry, but that might produce results, is if drug companies were encouraged to combine resources to combat particular disorders. The simple sharing of information about hits on targets might solve intractable problems, especially for diseases of the Third World such as tropical diseases, including African river blindness, sleeping sickness, and malaria. Cooperation over commercially interesting drugs is going to be market and bottom-line driven—like the developing co-marketing agreements and so on.

The recent moves to create registries of trial data are encouraging. In addition, a central repository for negative data about clinical candidates or targets that had been dropped would save the industry time and money.

If a target is dropped, it may be tainted forever throughout the industry, even though it is not generally revealed if it is actually a bad target or just a bad molecule, and patients will never benefit from this research on a putative drug target. If a company holds a patent on a target but drops the program without licensing the target, nothing can be done by others. Currently, companies can keep quiet about adverse reactions;[162] only when they have applied for an IND do they have to disclose adverse reactions to the FDA. In these cases, the FDA is aware of, but holds confidential, negative data, the sharing of which would be of industrywide utility. Of course, should negative datasets become available, there would be a need for some form of peer review. Bad negative data are bad data, just as bad positive data derived from poor experimental protocols in scientific research are for very good reasons not made publicly available in scientific literature. However, good experiments that fail have inherent informational value. It would be the same for negative trial data.

The problem is how to compensate for the time and money spent by one company, which now will be saved by the next one, giving a perceptively unfair competitive advantage in the event of a successful drug emerging. One may imagine—extrapolating from the way we share data in the public domain—that these data could become part of a semipublic domain, where companies could subscribe—pay some sums in relation to how many things they want to know of which type, and how many areas they want to hear about. Funds received, and perhaps extended rights, would partly go back to the depositor of the information—that is, the first company or even to the basic academic research entity, via the NIH. It is still hard to see how commercially acceptable it will be for the first company—because what it learned from the failed trial may have cost $50 to 300 million and it lost, in addition, time and opportunity. These kinds of sums are hard to recover from fees.

Another idea with which we toyed is based on the way companies pay royalty to universities for know-how and IP; the second and third companies who utilized these data would pay a royalty to the first one. The royalty rate would be balanced to make it advantageous to both to participate. In this case, a success-related fee would be given that is easier to see as just. We suspect, though, that many companies would not want to allow their competitor to succeed where they failed with the same data, because their company legend is that they are smarter than the competition.

---

[162] PCP was developed in a drug company and escaped to be a street drug: angel dust. Similarly, viral vectors as bio-warfare agents may have started in legitimate pharmaceutical research.

The companies do not have to justify why they drop candidates, the reasons for which might be almost whimsical—or at least based on spurious arguments—or because the company switched strategic thrust. How companies might be incentivized to provide data about their failures or dropped candidates that might benefit their competitors is difficult to imagine. But it is worth debating, if only to recognize and then hopefully avoid the unnecessary loss of intellectual and monetary expenditure when candidates are dropped because of rumors that a company's trial failed. Instead of keeping failures secret, disclosures such as these could be made while protecting their intellectual property interests.

The main hope, as mentioned earlier, is that licensing departments will turn failed programs into out-licensing opportunities.

Some Things Must Change

The future will be better served by changes in scientific approach to the problems of pharmaceutical disease management than by legislation. When drug companies have fully exploited the more traditional targets such as G-protein coupled receptors (GPCRs), the demand for product development will force them to find other approaches to improve therapeutic ratios. Some answers will come using traditional small-molecule pharmaceuticals; other problems will be solved by biologicals and their administration. But improving therapeutic ratios within clinical trials is hard. We hope that drugs in clinical practice will find their niches better than can be determined from trial data alone. This requires brave, well-educated physicians trying safe drugs for widespread indications.

The biggest change should eventually come from the implementation of pharmacogenomics-based individualized medicine. The theoretical basis for this approach is rapidly developing. The Human Genome Project has metamorphosed into the Haplotype Consortium.[163] The original human genome sequence belonged to no single individual, and, therefore, long or significant sequences do not belong to any individual "in the wild," as it were. The so-called Genetic Variation Mapping Project or "HapMap" will help identify genetic contributions to common disease. It is starting with the major histocompatability complex (MHC), genetic variations of

[163] See for example: http://www.nih.gov/news/pr/oct2002/nhgri-29.htm or http://www.the-scientist.com/yr2002/apr/cohen_p7_020415.html or http://www.sanger.ac.uk/HGP/Chr6/MHC/consortium.shtml

which have material consequences in the immune systems of the individuals carrying them. It is known that individuals in some population groups have predisposition or resistance to some autoimmune disorders such as multiple sclerosis or type I diabetes, and scientists will be able to link this attribute to particular sequences.

The estimate is that the human population can be separated according to genetic variations into 1,300 to 2,000 haplotypes—with Japan being the most homogeneous and Africa being the most heterogeneous population. Individuals in a group or a group of groups will respond differently to drug treatments. The future goal is to make this variation in response—whether it be a better therapeutic ratio or worse adverse reaction—predictable.

But the projected success of genomics-based individualized medicine is built on the assumption of huge changes in society's attitudes. Society is not yet ready for this, especially in a health environment built around a private insurer-based infrastructure wherein people with perceived or real genetic defects and their jobs are not automatically protected. In the increasingly rarified atmosphere of socialized medicine practiced in most of Europe, Japan, and Canada, where individuals are better protected, medicines targeted to the individual would be less problematic, though the cost might be restrictive, if not actually prohibitive. The biggest problem is that individuals projected to have increased health difficulties because of genetic disposition should not be penalized by increased job insecurity or elevated health insurance costs.

All these things must happen simultaneously; it is not a trivial thing.

Even if these problems can be solved by legislative action, the great benefit of genotyping must be demonstrated for patient, doctor, and insurer. Drug companies must make drugs for individuals or groups of individuals sharing specific genetic sequences. Smaller markets will be accompanied with smaller trials overseen by an FDA governed by different assumptions. The small trials will need more inventive statistical analysis. Some indications may be so small that one would need to enroll everyone on this planet who had the disease in the trial. While this is clearly absurd, it is possible—and is being done in Europe—to have all treating physicians in a trial connected by e-mail so that the physicians can share observations, whether positive or negative, during a trial in order to ensure more astute patient monitoring during the trial. The reporting of side effects will go through the FDA, and, furthermore, may be handled medically during the trial. This will help reestablish the physician–patient interaction prior to establishing the purely objective evidence-based medicine.

Bottom Lines

There are several bottom lines.

Pharmacogenomics will eventually meet pharmacoeconomics. There have been many, many strong examples of even expensive medicines being cost effective. Cholesterol-lowering medicines, at a cost of less than $3 a day, can help patients avoid coronary bypass surgery at a cost of about $75,000. Medicine that can stop the effects of osteoporosis, at a cost of about $730 a year, can prevent hip fractures, which cost an estimated $41,000 per patient. Preventing work loss through medical intervention is clearly of benefit. Medicines can have great fiscal value as well as by virtue of their increasing the quality of life.

The idea of the United States spending more on prescription drugs goes against the prevailing concerns of states and governors trying to meet budgets. However anything that drives down creative investment in new medicine development is going to have a negative effect on a society that can afford it.

Why has a society that has been so successful as the United States developed such a thirst for antidepressants and anxiolytics and at the same time as having almost constitutionally embraced capitalism has formed a powerful lobby objecting to the power and profits of especially Big Pharma?

Benign and creative leadership in the FDA, Big Pharma, and Biotech will make better drugs more available. An affluent society needs to be able to afford and enjoy the benefits of drug discovery.

Next?

There can always be more to say. By the time this book is published, the reader will know more about how porous the Canadian borders are to U.S. patients. All the time patients have to pay, the borders to cheaper drugs will always be porous we are guessing. But we cannot express this opinion without also referencing a concern that patients self-medicating by acquiring drugs from Canada or Mexico or wherever next is inherently bad, since it is extremely likely that a good number of drugs are being taken inappropriately, if not actually abused. Better insurance cures the border permeability.

What will be the big trends? Will Maine and Oregon represent the common bicoastal extreme of state interference in the pricing and budgets of prescription medicine, or will more and more states legislate according to this model?

For Big Pharma, and Mini- to Midi-Biotech, mergers are likely to continue, and more Mini-Biotechs will continue to start up as managements are released from merged entities. Large R&D organizations will be broken down into smaller R&D units of 200 to 300 people. Biotechs will continue to develop internal expertise and acquire medicinal chemistry departments. Big Pharma will continue to mimic Biotech by doing lots of biologicals. There is a danger that research is slowing since the number of people entering life sciences is not growing so fast. The Human Genome Project is done, but HapMap is happening.

One thing Big Pharma may not be able to combat is the development of the developing world. There is no, as yet, Big Pharma in the underdeveloped world, but the recent obvious entry of India and China may change Big Pharma ways as these countries manufacture more and increasingly research drugs. Brazil is to become an emerging market if it hasn't already. China is the future market. All Big Pharma is opening research laboratories there today. India and China have the market size, but not yet the money, although today India has as many families with high income as Germany. Food for thought, no doubt.

INDEX